Lewis Wright

The Practical Poultry Keeper

A Complete and Standard Guide to the Management of Poultry. Fourth Edition

Lewis Wright

The Practical Poultry Keeper
A Complete and Standard Guide to the Management of Poultry. Fourth Edition

ISBN/EAN: 9783337124090

Printed in Europe, USA, Canada, Australia, Japan

Cover: Foto ©Lupo / pixelio.de

More available books at **www.hansebooks.com**

SILVER-PENCILLED HAMBURGHS.

THE

PRACTICAL
POULTRY KEEPER:

A Complete and Standard Guide

TO THE

MANAGEMENT OF POULTRY,

WHETHER FOR

DOMESTIC USE, THE MARKETS, OR EXHIBITION.

By L. WRIGHT.

FOURTH EDITION.

NEW YORK:
THE ORANGE JUDD COMPANY,
245, BROADWAY.

By special arrangement, the sale of this book in the United States is placed in the hands of THE ORANGE JUDD COMPANY, of New York.

CASSELL PETTER & GALPIN,
London, Paris & New York.

CONTENTS.

SECTION I.

THE GENERAL MANAGEMENT OF DOMESTIC POULTRY, WITH A VIEW TO PROFIT:—

    PAGE

Chapter I.—Houses and Runs; and the Appliances necessary to keeping Poultry with Success 3

Chapter II.—On the System of Operations, and the Selection of Stock 14

Chapter III.—The Feeding and General Management of adult Fowls 20

Chapter IV.—Incubation 34

Chapter V.—The Rearing and Fattening of Chickens ... 44

Chapter VI.—Diseases of Poultry 55

SECTION II.

THE BREEDING AND EXHIBITION OF PRIZE POULTRY:—

Chapter VII.—Yards and Accommodation adapted for Breeding Prize Poultry 63

Chapter VIII.—On the Scientific Principles of Breeding, and the Effects of Crossing 70

Chapter IX.—On the Practical Selection and Care of Breeding Stock, and the Rearing of Chickens for Exhibition... 80

Chapter X.—On "Condition," and the Preparation of Fowls for Exhibition; and various other Matters connected with Shows 90

SECTION III.

DIFFERENT BREEDS OF FOWLS: THEIR CHARACTERISTIC POINTS, WITH A COMPARISON OF THEIR MERITS AND PRINCIPAL DEFECTS:—

Chapter XI.—Cochin-Chinas or Shanghaes 101

Chapter XII.—Brahma Pootras 105

	PAGE
Chapter XIII.—Malays	116
Chapter XIV.—Game	118
Chapter XV.—Dorkings	126
Chapter XVI.—Spanish	131
Chapter XVII.—Hamburghs	138
Chapter XVIII.—Polands	145
Chapter XIX.—French Breeds	151
Chapter XX.—Bantams	162
Chapter XXI.—The "Various" Class	167

SECTION IV.

TURKEYS, ORNAMENTAL POULTRY, AND WATERFOWL:—

Chapter XXII.—Turkeys. Guinea-fowl. Pea-fowl	173
Chapter XXIII.—Pheasants	185
Chapter XXIV.—Water-fowl	190

SECTION V.

THE HATCHING AND REARING OF CHICKENS ARTIFICIALLY:—

Chapter XXV.—The Incubator and its Management	203
Chapter XXVI.—Rearing Chickens Artificially	213

SECTION VI.

THE BREEDING AND MANAGEMENT OF POULTRY UPON A LARGE SCALE:—

Chapter XXVII.—Separate Establishments for Rearing Poultry. Poultry on the Farm. Conclusion ... 221

LIST OF ILLUSTRATIONS.

	PAGE
GROUND PLAN AND ELEVATION OF POULTRY-HOUSE	11
FEEDING-DISH	26
COVER FOR FEEDING-DISH	26
POULTRY-FOUNTAIN	29
PROTECTION FOR THE HEN WHILST SITTING	38
MODE OF TESTING EGGS	41
COOP FOR HEN WITH CHICKENS	46
COOP FOR TRANSPORTING HEN AND BROOD	48
WIRE-COVERED RUN FOR YOUNG CHICKENS	49
FATTENING PENS	52
PLAN OF MR. H. LANE'S POULTRY-YARD	65
" MR. R. W. BOYLE'S "	68
CRÈVECŒUR FOWLS	152
LA FLÈCHE COCKEREL	155
" PULLET	156
HOUDAN COCK	157
" HEN	158
BREDA FOWLS	159
SILKY "	169
BARN-DOOR "	171
BRINDLEY'S INCUBATOR	205
SECTION OF MR. F. H. SCHRÖDER'S INCUBATOR	207
MESSRS. GRAVES' INCUBATOR	212
MR. F. H. SCHRÖDER'S "MOTHER"	215
THE POULTRY-HOUSE AT BELAIR	225

LIST OF ILLUSTRATIONS.

	PAGE
VIEW OF HEN-HOUSE	226
PLAN OF ,,	227
THE OPEN RUNS FOR CHICKENS AND FOWLS	229
ARRANGEMENT OF GROUND FLOOR OF POULTRY-HOUSE	230
THE HATCHING-ROOM	231
FEEDING-COOPS	232
THE PORTABLE HATCHING BASKET	233
THE KITCHEN	233
THE STORE-ROOM	234
THE AUDEOD CORN-BIN	235

SEPARATE PLATES.

SILVER-PENCILLED HAMBURGHS		*Frontispiece*
WHITE COCHINS	Facing p.	101
FEATHERS	,,	107
DARK BRAHMAS	,,	108
DUCK-WING GAME	,,	121
GREY DORKINGS	,,	127
WHITE-FACED BLACK SPANISH	,,	131
SILVER-SPANGLED POLANDS	,,	147
BLACK AND SEBRIGHT BANTAMS	,,	163
VARIEGATED CAMBRIDGE TURKEYS	,,	180
ROUEN AND AYLESBURY DUCKS	,,	191
TOULOUSE GEESE	,,	197

PREFACE.

WITH at least half a dozen books upon the subject of Poultry already at the choice of the reader, some apology may be deemed necessary for the publication of yet another.

Such our apology is very brief, and rests in the simple fact that a *practical* treatise—authoritative and comprehensive, yet simple and popular—has yet to be supplied. We are not to the present time aware of any work we could put into the hands of a person totally *ignorant* of poultry keeping, with the reasonable certainty that its instructions, if followed, would command success. Descriptions of breeds there are in plenty, some of them of great value; but very little has been written respecting the practical details of even ordinary poultry management; and with regard to two very important parts of the subject—the breeding and rearing of poultry for exhibition, and artificial incubation—absolutely nothing has yet been published in a connected form.

To occupy this field is one object of the following pages; which are the fruit of a thorough practical experience and knowledge of fowls, and will, we believe, be found a plain and sufficient guide to the merest tyro in any circumstances that are likely to occur to him; whilst even experienced breeders, we hope, may also find hints which may be useful to them.

Yet, whilst thus paying special attention to practical management, the different breeds have not been overlooked; and of every leading variety, at least, sufficient description has been given to answer every purpose of the fancier. Where we have permission to give them, the eminent names appended to the different chapters will be ample guarantee for the correctness of this portion of our work; but it is in every case to be understood that we do not rely alone upon our own careful study of the best specimens, but have the highest authority in each breed for every statement made respecting it. It is the more necessary to state this, because we have been compelled in a few cases to dissent from the well-known "Standard of Excellence"—usually most reliable, and hitherto the generally recognised authority on this part of the subject.

It only remains to thank those who have aided us, and placed their valuable experience and knowledge at the service of the public. Some of these are old friends : others are, or at least were, personal strangers. But in either case we feel pleasure in recording that, in nearly every instance, any assistance requested has been accorded as frankly as it was asked, and has frequently led to after intercourse of a most pleasant kind; and that to the cheerful and kindly aid of the most eminent breeders in the kingdom these pages owe much of whatever value they may have.

Kingsdown, Bristol,
January 31, 1867.

SECTION I.

THE

GENERAL MANAGEMENT OF DOMESTIC POULTRY
WITH A VIEW TO PROFIT.

TO THE READER.

The pages of this Section are not intended simply to be read and commended; but the directions given are such as are proper for the circumstances therein referred to, and are the price to be paid for health and eggs.

For instance: when it is said that the roosting-house should be cleansed daily, it is meant that it *should be done.* When it is said that fowls in confinement should have daily fresh vegetable food, it is intended to convey that such food must be *regularly given.* And so on.

Let the reader deal fairly by us and by his poultry. So will the latter deal fairly by him.

GENERAL MANAGEMENT OF FOWLS.

CHAPTER I.

HOUSES AND RUNS; AND THE APPLIANCES NECESSARY TO KEEP-
ING POULTRY WITH SUCCESS.

FOWLS should not be kept unless proper and regular attention can be given to them; and we would strongly urge that this needful attention should be *personal*. Our own experience has taught us that domestics are rarely to be relied upon in many matters essential both to economy and the well-being of the stock; and, if any objection be made on the score of dignity, we could not only point to high-born ladies who do not think it beneath them to attend to their own fowls, but can aver that even the most menial offices can be performed in any properly-constructed fowl-house without so much as soiling the fingers. If there be children in the family old enough to undertake such matters, they will be both pleased and benefited by attending to what will soon become their pets; if not, the owner must either attend to them himself, or take such oversight as shall be *effectual* in securing not only proper care of his birds, but of his own meal and grain. If he be unable or unwilling to do at least as much as this, he had far better not engage in such an undertaking at all.

The first essential requisite to success in poultry-keeping is a *thoroughly good* house for the birds to roost and lay in. This does not necessarily imply a large one or a costly: we once knew a young man who kept fowls most profitably, with only a house of his own construction not more than three feet square, and a run of the same width, under twelve feet long. It means simply that the fowl-house must combine two absolute essentials—be both perfectly weatherproof, and well ventilated.

With regard to the first point, it is not only necessary to keep out the rain but also the *wind*—a matter very seldom attended to as it ought to be, but which has great influence on the health and laying of the inmates. The cheapest material is wood, of which an inch thick will answer very well in any ordinary English climate; but if so built, the boards must either be tongued together, or all the cracks between them carefully caulked by driving in string with a blunt chisel. Care should also be taken that the door fits well, admitting no air except under the bottom; and, in short, every precaution taken to prevent draught. The hole by which the fowls enter, even when its loose trap-door is closed, should admit enough air to supply the inmates, and the object is to have but this *one* source of supply, and to keep the fowls out of all direct draught from it. For the roof, tiles alone are not sufficient, and if employed at all, there should be either boarding or ceiling under them; otherwise all the heat will escape through the numerous interstices, and in winter it will be impossible to keep the house warm. Planks alone make a good roofing. They may either be laid horizontally, one plank overlapping the other, and the whole well tarred two or three times first of all, and every autumn afterwards; or perpendicularly, fitting close edge to edge, and tarred, then covered with large sheets of brown paper, which should receive two coats of tar more. This last makes a very smooth, weatherproof, and durable roofing, which throws off the water well. But, on the whole, we prefer board covered with patent felt, which should be tarred once a year.

In the north of England, a house built of wood, unless artificially warmed, requires some sort of lining. Matting is often used, and answers perfectly for warmth, but unfortunately makes a capital harbour for vermin. If employed at all, it should only be slightly affixed to the walls, and at frequent intervals be removed and well beaten. Felt is the best material, the strong smell of tar repelling most insects from taking up their residence therein.

If a tight brick shed offers, it will, of course, be secured for the poultry habitation. But let all dilapidations be well repaired.

Ventilation is scarcely ever provided for as it should be, and the want of it is a fruitful source of failure and disease. An ill-ventilated fowl-house *must* cause sickly inmates; and such will never repay the proprietor. This great desideratum must, however, as already observed, be secured without exposing the fowls to any direct draught; and for the ordinary detached fowl-houses, the best plan is to have an opening at the highest point of the roof, surmounted by a "lantern" of boards, put together in the well-known fashion of Venetian blinds.

A south or south-east aspect is desirable, where it can be had; and to have the house at the back either of a fire-place or a stable is a *great* advantage in winter; but we have proved by long experience that both can be successfully dispensed with if only the two essentials are combined, of good ventilation with perfect shelter.

We do not approve of too large a house. For half-a-dozen fowls, a very good size is five feet square, and sloping from six to eight feet high. The nests may then be placed on the ground at the back, where any eggs can be readily seen; and one perch will roost all the birds. This perch, unless the breed kept is small, had better not be more than eighteen inches from the ground, and should be about four inches in diameter. A rough pole with the bark on answers best: the claws cling to it nicely, and bark is not so hard as planed wood. By far the greater number of perches are much too high and small; the one fault causing heavy fowls to lame themselves in flying down, and the other producing deformed breastbones in the chickens—an occurrence disgraceful to any poultry-yard. The air at the top of any room or house is, moreover, much more impure than that nearer the floor.

Many prefer a movable perch fixed on trestles. In large houses they are useful, but in a smaller they are needless. If the perch be placed at the height indicated, and a little in advance of the front edge of the nests, placed at the back, no hen-ladder will be required; and the floor being left quite clear, will be cleaned with the greatest ease, while the fowls will feel no draught from the door.

Besides the house for roosting and laying, a shed is necessary, to which the birds may resort in rainy weather. Should the house, indeed, be very large and have a good window, this is not *absolutely* needed; otherwise it must be provided, and is better separate in any case. If this shed be fenced in with wire, so that the fowls may be strictly confined during wet weather, so much the better; for next to bad air, wet is by far the most fruitful source, not only of barrenness, but of illness and death in the poultry-yard. If the space available be very limited—say five or six feet by twelve or sixteen—the whole should be roofed over; when the house will occupy one end of the space, and the rest will form a covered "run." But in this case the shed should be so arranged that *sun-light* may reach the birds during some part of the day. They not only enjoy it, but without it, although adult fowls may be kept for a time in tolerable health, they droop sooner or later, and it is almost impossible to rear healthy chickens.

Should the range be wider, a shed from six to twenty feet long and four to eight wide may be reared against the wall. Next the fowl-house will still, for obvious reasons, be the most convenient arrangement, and it is also best fenced in, as before recommended. The whole roof should be in one to look neat, and should project about a foot beyond the enclosed space, to throw the water well off. To save the roof drippings from splashing in, a gutter-shoot will of course be provided, and the wire should be boarded up a foot from the ground. All this being carried out properly, the covered "run" ought at all times to be perfectly dry.

The best flooring for the fowl-house is concrete made with strong, fresh-slaked hydraulic lime and pounded "clinkers," put down hot, well trodden once a day for a week, and finally smoothed. The process is troublesome, but the result is a floor which is not only very clean in itself, but easily *kept* so. Trodden earth will also answer very well. The floor of the shed may be the same, but, on the whole, it is preferable there to leave the natural loose earth, which the fowls delight to scratch in.

Cleanliness *must* be attended to. In the house it is easily secured by laying a board under the perch, which can be scraped clean every morning in a moment, and the air the fowls breathe thus kept perfectly pure. Or the droppings may be taken up daily with a small hoe and a housemaid's common dustpan, after which a handful of ashes or sand lightly sprinkled will make the house all it should be.

There is another most excellent plan for preserving cleanliness in the roosting-house, for which we are indebted to *The Canada Farmer*, and which is shown in Fig. 1. A broad shelf (*a*) is fixed at the back of the house, and the perch placed four or five inches above it, a foot from the wall. The nests are conveniently placed on the ground underneath, and need no top, whilst they are perfectly protected from defilement and are also well shaded, to the great delight of the hen. The shelf is scraped clean every morning with the greatest ease and comfort, on account of its convenient height, and slightly sanded afterwards; whilst the floor of the house is never polluted at all by the roosting birds. The broad shelf has yet another recommendation in the perfect protection it affords from upward draughts of air.

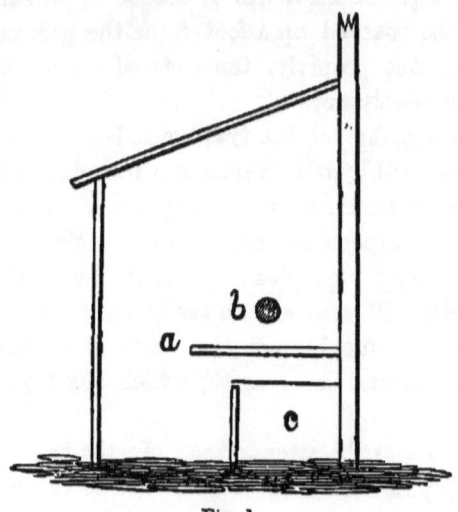

Fig. 1.

a Broad shelf, eighteen inches high.
b Perch, four inches above.
c Nests, open at top and in front.

The covered "run" should be raked over two or three times a week, and *dug* over whenever it looks sodden or gives any offensive smell. Even this is not sufficient. Three or four times a year, two or three inches deep—in fact, the whole polluted soil—must be removed, and replaced by fresh earth, gravel, or ashes, as the case may be.

Under the shed must be constantly kept a heap of dry dust or sifted ashes, for the fowls to roll in and cleanse themselves in their own peculiar manner, which should be renewed as often as it becomes damp or foul from use.

If chickens be a part of the intended plan, a separate compartment should be provided for the sitting hens; but this will be further treated of in a subsequent chapter.

Many will wish to know what space is *necessary*. The "run" for the fowls should certainly be as large as can be afforded; an extensive range is not only better for their health, but saves both trouble and food, as they will to a great extent forage for themselves. Very few, however, can command this; and poultry may be kept almost anywhere by bearing in mind the one important point, that the smaller the space in which they are confined, the greater and more constant attention must be bestowed upon the cleanliness of their domain. They decline rapidly in health and produce if kept on foul ground. If daily attention be given to this matter, a covered shed ten or twelve feet long by six feet wide, may be made to suffice for half a dozen fowls without any open run at all. By employing a layer of dry earth as a deodoriser, which is turned over every day and renewed once a week, the National Poultry Company kept for several years such a family in each pen of their large establishment at Bromley. These pens did not exceed the size mentioned, yet the adult fowls at least were in the highest health and condition; and the company managed, with birds thus confined, to take many prizes at first-class shows.

Poultry-keeping is, therefore, within the reach of all. The great thing is purity, which *must* be secured, either by space, or in default of that, by care. Hardy fowls will sometimes thrive in spite of draughts, exposure, and scanty food; but the strongest birds speedily succumb to bad management in this particular, which is perhaps the most frequent cause of failure. It should also be remarked that poultry thus confined will require a different diet to those kept more at liberty; but this will be more fully explained in a succeeding chapter.

If the run be on the limited scale described, dry earth is decidedly the best deodoriser. It is, however, seldom at the command of those who have little space to spare, and sifted ashes an inch deep, spread over the floor of the whole shed, will answer very well. The ashes should be raked every other morning, and renewed at least every fortnight, or oftener if possible. Of course, the number of fowls must be limited; they should not exceed five or six, and unless a second shed of the same size can be allowed, the rearing of chickens should not be attempted.

To those who can give up a portion of their garden, the following plan of a poultry-yard can be confidently recommended. It represents, with very slight modification, our own present accommodation; and having tested it by experience, we are prepared to say that it is not only more convenient, more simple, and more cheaply erected than any plan on a similar scale we have seen, but, with the addition of a lawn on which the chickens may be cooped, is adapted to rearing in the highest perfection any single variety of either ordinary or "fancy" fowls. The space required in all is only twenty-five by thirty-five feet. If more can be afforded, give it, by all means; but we have found this, with very moderate care, amply sufficient, and we believe it will meet the requirements of a larger class of readers than any other we are acquainted with.

The plan here given, it will be seen, comprises two distinct houses, sheds, and runs, with a separate compartment for sitting hens. The nests are placed on the ground at the back of the houses, and the perches, as before recommended, a foot in advance of them, and eighteen inches high. The holes by which the fowls enter open into the sheds, which are netted in, so that in wet weather they can be altogether confined. In dry weather the shed is opened to give them liberty. The fencing should be boarded up a foot high, not only to prevent rain splashing in, but to keep in when necessary young chickens, which would otherwise run out between the meshes.

A walk in front of the sheds should be gravelled, and the

ELEVATION

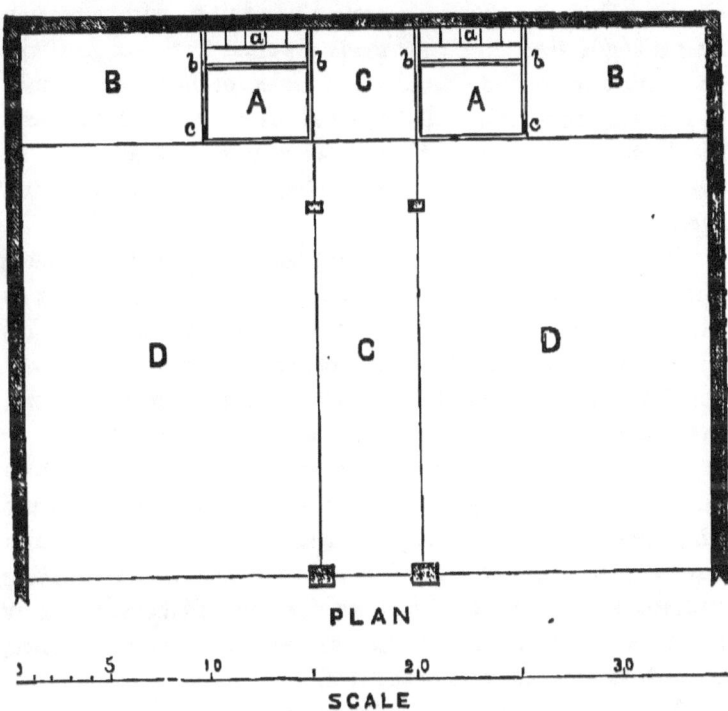

PLAN

SCALE

Fig. 2.

A A Roosting and laying houses.   a a Nests.
B B Fenced-in covered runs.   b b Perches.
C C Shed and run for sitting hens.   c c Holes for fowls to enter.
D D Grass runs.

remainder of the open runs laid down in grass, which, if well rooted first, will bear small fowls upon it for several hours each day, but should be renewed in the spring by sowing when needed. The runs should be enclosed with wire netting, two inches mesh, which may be conveniently stretched on poles $1\frac{1}{2}$ inch square, driven two feet into the ground, and placed five feet apart. The height of the fence depends on the breed chosen. Cochins or Brahmas are easily retained within bounds by netting a yard high; for moderate-sized fowls six feet will do; whilst to confine Game, Hamburghs, or Bantams, a fence of eight or nine feet will be found necessary. The netting should be simply stretched from post to post, without a rail at the top, as the inmates are then far less likely to attempt flying over.

We do not like to see fowls with their wings cut. If their erratic propensities are troublesome, open one wing, and *pluck out* all the first or flight feathers, usually ten in number. This will effectually prevent the birds from flying, and as the primary quills are always tucked under the others when not in use, there is no external sign of the operation.

The holes by which the fowls enter the houses should be furnished with trap doors, that they may be kept out at pleasure whilst either part is being cleaned. Each house must also have a small window. Having a shed at the side, ventilating lanterns will not be necessary, as the end will be attained by boring a few holes in the wall between the house and shed, towards the highest part of the roof.

The compartment for the sitting hen may be walled in at the front or not; for ourselves, we prefer it open. Her run may also be covered over or not, at pleasure. To have it in the middle, as here shown, we consider most convenient; but in our own case this compartment is at the *side*, instead of between the two houses, which we built close together. This was rendered advisable on account of our shed having unavoidably a due easterly exposure; and by so arranging the premises that each inhabited house should have one adjoining, we ensured to both one comparatively *warm side*, and thus, in a measure, counteracted the evil. We give this little bit of personal experience in order that the reader may see the way in which varying considerations are to be weighed before a plan is finally determined on.

Such a yard possesses many advantages. Two separate runs are almost necessary if the rearing of chickens forms part of the plan of proceeding. It is also in some respects convenient to keep two different breeds, as one may supply the deficiencies of the other; and many persons consider it advisable to separate the cocks and hens, except during the breeding season, believing that stronger chickens are obtained thereby. The need of the separate compartment for the sitting hens is further insisted on hereafter, but it has also other uses; being, when not so employed, often very convenient for the temporary reception of a pen of strange birds, for which there may be no other accommodation.

Each run will accommodate from six to ten fowls, according to their size and habits.

For those who purpose to engage in wholesale or prize poultry-breeding, more extensive designs will be given hereafter; but enough has now been said to enable the intending poultry-keeper to select from the different plans here indicated the one best adapted to his particular situation, or, mayhap, to contrive a better one of his own. We have pointed out the essentials; and these being provided for, operations can be commenced, and it becomes necessary to determine upon the plan of proceeding. This, however, will be more fully treated of in the next chapter.

CHAPTER II.

ON THE SYSTEM OF OPERATIONS, AND THE SELECTION OF STOCK.

WHEN poultry are kept as a branch of domestic economics, it will be obvious that the system to be pursued should vary according to the extent of accommodation which can be afforded, and to the object sought. Both these considerations should be well weighed before operations are commenced; and the plan then determined upon as best adapted to the circumstances should, as long as those circumstances remain the same, be consistently carried out and adhered to.

It very frequently happens that a regular supply of eggs is the sole object in view, and that neither the time, trouble, nor space required to rear chickens with success can well be spared. If, for instance, a covered shed fenced in with wire, as described in the last chapter, with a small house at the end for roosting and laying in, be the sole accommodation for the fowls, to attempt *rearing* them would be folly;* and yet they may be *kept* so as to yield a good return upon their cost and maintenance. The proper plan in such a case will be to purchase in the spring a number of hens proportioned to the size of the run, and none exceeding a year or eighteen months old. A cock is useless; as hens lay, if there be any difference, rather better without one, and where eggs only are wanted, his food is thrown away. All these birds, if in good health and condition, will either be already laying, or will commence almost immediately; and if well housed, as in the last chapter, and properly fed, will ensure a constant supply of eggs until the autumnal moulting season. Whenever a hen shows any desire to sit, the propensity must of course be checked—not by the

---

* It is not meant to be denied that chickens *can* be reared in such circumstances, and that in good health and to a fair size. We have ourselves done so; but it does not *pay*, and we do not intend to do it again.

barbarous expedient of half drowning the poor bird in cold water—a process generally as ineffectual as it is cruel, but, having allowed it to sit on the nest for four or five days, by shutting it up in a dark place, with plenty of water, but rather scanty food. The best plan is to invert a small cask, of which the head has been removed, upon three bricks. A hole being bored near the top for ventilation, this will make a capital pen for a "broody" hen, the food and water being placed just under the rim. A few days of such confinement will take away all desire to sit from almost any hens but Cochins, which should not be kept, on that account, under the circumstances we are considering; and in about a fortnight the fowl, if not older than we have recommended, will begin to lay again.

To buy only young and healthy birds is very important. An experienced hand can tell an old fowl at a glance, but it is rather difficult to impart this knowledge to a beginner, for no one sign is infallible, at least to an uninitiated interpreter. In general, however, it may be said that the legs of a young hen look delicate and smooth, her comb and wattles soft and *fresh*, and her general outline, even in good condition (unless fattened for the table), rather light and graceful; whilst an old one will have rather hard, horny-looking shanks, her comb and wattles look somewhat harder, drier, and more "scurfy," and her figure is well filled out. But any of these indications may be deceptive, and the only advice we can give the reader is, to use his own powers of observation, and try and catch the "*old look*." He will soon do so, and need no further description.

Directly these hens stop laying in the autumn, and before they have lost condition by moulting, they should, unless Hamburghs or Brahmas, be either killed or sold off, and replaced by pullets hatched in March or April, which will have moulted early. These again, still supposing proper food and good housing, will all be producing eggs by November at furthest, and continue, more or less, till the February or March following. They may then either be disposed of, and replaced as before, which we should ourselves prefer, as they are just in prime condition for the table; or, as they will not stop laying very long, the best of them may be retained till the autumn, when they *must* be got rid of.* For if fowls be kept for eggs it is essential to success that *every autumn* the stock be replaced with pullets hatched early in the spring. By *no other* means can eggs at this season be relied upon, and the poultry-keeper must remember that it is the *winter which determines* whether he shall gain or lose by his stock; in summer, if only kept moderately clean, hens will pay for themselves treated almost anyhow. The only exception to this rule is in the case of Cochins, Brahmas, or Hamburghs, which will lay through the winter up to their second, or even third year.

The stock to be selected, if a pure strain be chosen, are Hamburgh or Spanish; either, in favourable circumstances, will give a plentiful supply of eggs, and give no trouble on the score of sitting propensities. The Spanish lays five or six very large eggs a week in spring and summer, but is not a hardy or free-laying breed for winter, and must have a warm aspect and perfect shelter from wind, if the supply is to be kept up. Hamburghs are tolerably hardy, and are capital winter layers; they also produce more eggs *in a year* than any other breed, laying almost every day except when moulting, and never wanting to sit; but the eggs are rather small. More than four or five Hamburghs should not be put in a shed, and they must be kept *scrupulously* clean; with these conditions they will thrive, but few breeds suffer so much from filth or overcrowding. Brahmas may also be strongly recommended. As layers, they are in the very first class; are very tame, and bear confinement well; and the tendency to sit does not

* That is, if the greatest amount of profit be the object sought. The question of "pets," and the pleasure to be derived from them, we are not considering.

occur often enough to be troublesome, as in the case of Cochins.

When there is a good wide range of any kind, nothing will be so profitable as a few Game hens, the black-breasted red variety being best. The hens are as prolific as *any* breed whatever, and eat very little in proportion; but they cannot be kept in close confinement on account of their fighting propensities.

For ourselves, we prefer pure breeds, or first crosses; for after all is said on the superiority of mongrel fowls, where is the "barn-door" bird that will lay as many eggs as a Brahma or a Hamburgh? Still, the cost of a good stock will stand in the way with many, and has to be taken into consideration; and to those who cannot afford "fancy" poultry, it may therefore be said once for all, that on the whole, equal success may be attained with ordinary or "barn-door" fowls. Care must be taken in the selection. They should be young, fair-sized, sprightly-looking birds, with plump, full breasts, rather short legs, and nice tight-looking plumage, after such a type as shown on page 171; they ought also to be chosen from a country yard, where their *parents* have been well fed. If such be obtained, they will repay the purchaser, and are handsomer and better every way than *inferior* birds of the "fancy" class. Of course, this remark does not apply to mere faults of colour. Fowls are often to be met with at a moderate price, which from some irregularity of feather are quite disqualified as show birds, but which possess in perfection all the other merits of the breed to which they belong. Let such be secured and prized by all means; but let it be also remembered and believed, that nothing pays so wretchedly as to begin "poultry-fancying" with inferior stock, and that really fine fowls which never had a grandfather are any day preferable to "degenerate descendants from a line of kings."

It has been already remarked that the Cochin breeds are

excellent layers in winter, but that their invincible propensity to sit, which occurs every two months, or even less, is a fatal objection to their being kept by those who do not desire the care of young broods. If, however, the system adopted depend upon home-reared chickens to replenish the stock, one or two Cochin hens may be kept with great advantage, especially if the other fowls are Spanish or Hamburgh. The frequency of their desire to incubate now becomes a recommendation, as the owner can depend upon "a broody hen" at almost any season which may suit his views; and if always parted with at the age of two years, they will not fail to maintain their deserved character as good winter layers. The number of such hens must depend upon circumstances. If it be only intended to replace from time to time the laying stock, or to hatch the eggs of non-sitting varieties, one or two Cochins will furnish more broods than will be required; and when their services are no longer needed in this way, the desire to sit must be hindered as already described. In this case the eggs should be set in March or April, that the young pullets may begin to lay early. In proportion to the number of broods desired may the number of Cochins be increased; and if a constant supply of chickens for the table be—as it often is—the main end in view, they may form a very considerable portion of the stock, and every hen may be set in turn. Their own eggs, of course, should not be given them if the chickens be for market, unless running with a Dorking, Houdan, or Crèvecœur cock, either of which crosses produces a gigantic table-fowl of very fair edible qualities. For *home* use, however, Cochins are not to be despised when killed anywhere under nine months old; they carry an immense quantity of solid meat; and if this be more in the leg than could be desired, it must be also remembered that the said leg, though certainly *not* equal to breast or wing, is more tender than that of most other breeds.

On the whole, however, if a good stock can be afforded,

and one or two broods of chickens yearly can be managed, we should, for domestic use, recommend Dark Brahmas. The light variety is also good, but the dark is the larger fowl, and looks best under confinement. If there be a double run, as described in Chapter I., the finest birds may be kept pure, and their eggs and progeny, when possible, sold at "fancy" prices; whilst the hens which show faults of colour may be kept in the other run with a large coloured Dorking or Crèvecœur cock. From this cross table-fowls may be obtained which "look like young turkeys," and being hardy are easily reared. The flesh may not be quite equal to that of the game fowl—in delicious flavour "the prince of all breeds"—but it equals the Dorking, with greater size, and freedom from that very delicate constitution which often renders the latter an unprofitable fowl.

Dorkings, notwithstanding, are not to be despised, and will do well if they have a fair-sized run, well gravelled, and free from wet, with a good dry shed to shelter in. If the supply of table poultry be a main point, no breed, except perhaps Houdans, will compare with this, the favourite fowl of the London market. When of good stock, they may be got up to an amazing size, and the quality of the meat is excellent. They are also most exemplary mothers, and in moderate weather produce a very fair quantity of eggs; but are not very good winter layers, even when hatched early. In this respect they are excelled by the recently-introduced Houdans, which lay very freely, and are also most hardy fowls, whilst in size and quality of flesh they equal the Dorking, whose blood, though perhaps generations back, we believe them to share, as evidenced by the general form and the peculiar fifth toe. We consider Houdans pre-eminently the breed for the farmer. They will ultimately be bred larger than even Dorkings, which they equal now; and their extreme hardiness, quick growth, and excellent laying, give us a fowl with nearly all the excellences and but little of the faults of the fine old English breed.

On the whole, therefore, of the pure breeds, we should pronounce Houdans to be the *farmer's*, and Brahmas the *family* fowl, crossing the table-chickens from the latter with Dorking or not, according as there were one or two runs to keep them in. If a few eggs daily be the object, our own choice would be four or five spangled Hamburghs, provided there be a moderate run, or even a good-sized shed, and they be kept scrupulously clean and well sheltered from driving wind or rain. If the space be *very* limited, and economy be important, we would select four or five red-faced Spanish, or, as they are now called, Minorcas; they lay at least as well as their celebrated white-faced cousins, while they are far hardier in winter, and stand confinement very well; their price, also, being often very little more than that of common hens. In default of either of these, however, and if all be beyond the means of the speculator, we would undertake to show a satisfactory balance-sheet with any good, lively ordinary fowls.

Let us, however, repeat again—for nothing is so important—whatever be the breed selected, there must be *every autumn* a proportion, at least, *regularly replaced* by young birds hatched in the spring of the same year. This is the great secret of success, as far as system is concerned; and if it be neglected, during winter an empty egg-basket will eat up all the summer's profits, and testify dismally to the improvidence of the owner.

CHAPTER III.

THE FEEDING AND GENERAL MANAGEMENT OF ADULT FOWLS.

A JUDICIOUS system of feeding is very essential to the well-being of poultry, and has, of course, more *direct* influence upon the profit or loss than any of the circumstances—though equally important—which we have hitherto enumerated. We shall, therefore, endeavour to give the subject a full, practical consideration.

The object is to give the quantity and quality of food which will produce the greatest amount of flesh and eggs; and if it be attained, the domestic fowl is unquestionably the most profitable of all live stock. But the problem is rather a nice one, for there is no "mistake on the right side" here. A *fat* hen is not only subject to many diseases, but ceases to lay, or nearly so, and becomes a mere drag on the concern; while a pampered male bird is lazy and useless at best, and very probably, when the proprietor most requires his services, may be attacked by apoplexy and drop down dead.

That fowls cannot be remunerative if starved need scarcely be proved. *Ex nihilo nihil fit;* and the almost daily production of an article so rich in nitrogen as an egg—the very essence of animal nourishment—*must* demand an ample and regular supply of adequate food. We say no more upon this point, knowing that the common mistake of nearly all amateur poultry-keepers is upon the other side—that of over-feeding.

The usual plan, where fowls are regularly fed at all, appears to be to give the birds at each meal as much barley or oats as they will eat; and this being done, the owner prides himself upon his liberality, and insists that *his* at least are properly fed. Yet both in quantity and quality is he mistaken. Grain will do for the regular meals of fowls which live on a farm, or have any other extensive range where they can provide other food for themselves, have abundant exercise, and their digestive organs are kept in vigorous action. But poultry kept in confinement on such a diet will not thrive. Their plumage, after a while, begins to fall off, their bowels become affected, and they lose greatly in condition; and though in summer their eggs may possibly repay the food expended, it will be almost impossible to obtain any in winter, when they are most valuable.

Even those who profess to correct such errors are not always safe guides. We have before us a work that stands high both in character and price, and is in many respects really valuable, in which, just after a caution against overfeeding, the editor gives five pounds of barley meal, ten pounds of potatoes, seven pounds of oats, three pounds of rice boiled, and three pounds of scalded bran, as a week's allowance for five hens and a cock—"of the larger kinds" it is true. Now, at the lowest ordinary prices the cost of such a scale would amount to, at least, £4 4s. in the course of twelve months; and taking eggs at the high average of a penny each all the year through, every one of the five hens must lay, at least, 200 eggs to repay the mere cost of their subsistence. When we say that 150 eggs per annum is as much as can be obtained from nine hens out of ten, it will be seen at once that poultry *could* not be made profitable did they consume so enormously; and, in point of fact, we had the curiosity to try this dietary upon six fowls "of the larger kinds," and found it rather more than double what was amply sufficient.

The fact is, all fixed scales are delusive. Not only would Cochins or Crèvecœurs eat twice as much as many other sorts; but different fowls of the same breed often have very different measures of capacity, and even the same hen will eat nearly twice as much while in active laying as when her egg-organs are unproductive.

The one simple rule with adult fowls is, to give them as much as they will eat *eagerly*, and no more; directly they begin to feed with apparent indifference, or cease to *run* when the food is thrown at a little distance, the supply should be stopped. In a state of nature, they have to seek far and wide for the scanty morsels which form their subsistence; and the Creator never intended that they, any more than human beings, should eat till they can literally eat no more. It follows, from this rule, that food should never be left on the ground. If such a slovenly practice be permitted, much of what is eaten will be wasted, and a great deal will never be eaten at all; for fowls are dainty in their way, and unless at starvation point always refuse sour or sodden food.

The number of meals per day best consistent with *real* economy will vary from two to three, according to the size of the run. If it be of moderate extent, so that they can, in any degree, forage for themselves, two are quite sufficient, at least in summer, and should be given early in the morning, and the last thing before the birds go to roost. In any case, these will be the principal meals; but when the fowls are kept in confinement, they will require, in addition, a *scanty* feed at mid-day.

The first feeding should consist of *soft food* of some kind. The birds have passed a whole night since they were last fed; and it is important, especially in cold weather, that a fresh supply should as soon as possible be got into the *system*, and not merely into the crop. But if grain be given, it has to be *ground* in the poor bird's gizzard before it can be digested; and on a cold winter's morning the delay is anything but beneficial. But for the very same reason, at the *evening* meal grain forms the best food which can be supplied; it is digested slowly, and during the long cold nights affords support and warmth to the fowls.

A great deal depends upon this system of feeding, which we are aware is opposed to the practice of many, who give grain for the breakfast, and meal, if at all, at night. We believe such a system to be usually adopted from indolence; it is easier to throw down dry grain in a winter's morning than to properly prepare a feed of meal, which is accordingly given at night instead. Fowls so treated, however, are much more subject to roup and other diseases caused by inclement weather than those fed upon the system we recommend—a system not only in accordance with theory and our own experience, but with that of the most successful breeders. Let the sceptical reader make one simple experiment. Give the fowls a feed of meal, say at five o'clock in the evening; at twelve visit the roosts, and feel the crops of the poor birds. All will be empty; the gizzard has nothing to act upon, and the food speedily disappears, leaving with an empty stomach, to cope with the long cold hours before dawn, the most hungry and incessant feeder of all God's creatures; but if the last feed has been grain, the crop will still be found partially full, and the birds will awake in the morning hearty, strengthened, and refreshed.

With respect to the morning meal of pultaceous food, when only a few fowls are kept, to supply eggs for a moderate family, this may be provided almost for nothing by boiling daily the potato peelings till soft, and mashing them up with enough bran, slightly scalded, to make a tolerably stiff and dry paste. There will be more than sufficient of this if the fowls kept do not exceed one for each member of the household; and as the peelings cost nothing, and the bran very little, one half the food is provided at a merely nominal expense, while no better could be given. A little salt should always be added, and in winter a slight seasoning of pepper will tend to keep the hens in good health and laying. This food may be mixed boiling hot over night, and covered with a cloth, or be put in the oven; in either case it will remain warm till morning—the condition in which it should always be given in cold weather.

If a tolerable stock of poultry be kept, such a source of supply will be obviously inadequate; and in purchasing the food there is much variety to choose from. Small or "pig" potatoes may be bought at a low price and similarly treated; or barley-meal may be mixed with hot water; or an equal mixture of meal and "sharps," or of Indian meal and bran; either of these make a capital food. Or, if offered on reasonable terms, a cart-load of swede or other turnips, or mangel-wurtzel, may be purchased; and when boiled and mashed with meal or "sharps," we believe forms the *very best* soft food a fowl can have, especially for Dorkings; but they cannot everywhere be obtained at a cheap rate, and the buyer must study the local market.

A change of food, at times, will be beneficial, and in making it the poultry-keeper should be guided by the season. When the weather is warm, and the production of eggs abundant, the food should abound in nitrogenous or flesh-forming material, and not contain too much starch or oil, both of which, being carbonaceous, have warmth-giving and fattening properties; but when the cold weather approaches, and the eggs even of good winter layers are fewer than in summer, less of nitrogenous and more of carbonaceous food will be needed. The following table has been often copied since its first publication in the "Poultry Diary;" but its practical usefulness is so obvious that we make no apology for giving it here, with some modification to make the proportion of warmth-giving to flesh-forming ingredients more plain.

There is in every 100 lbs. of	Flesh-forming Food.	Warmth-giving Food.		Bone-making Food.	Husk or Fibre.	Water.
	Gluten, &c.	Fat or Oil.	Starch, &c.	Mineral Substance.		
Oats	15	6	47	2	20	10
Oatmeal	18	6	63	2	2	9
Middlings or fine Sharps	18	6	53	5	4	14
Wheat	12	3	70	2	1	12
Barley	11	2	60	2	14	11
Indian Corn...	11	8	65	1	5	10
Rice	7	A trace.	80	A trace	—	13
Beans & Peas	25	2	48	2	8	15
Milk	4½	3	5	¾	—	86¾

To show the practical use of this table, it may be observed that whilst "sharps" or "middlings," from its flesh-forming material, is one of the best summer ingredients, in winter it may be advantageous to change it for a portion of Indian meal.

It is, however, necessary to avoid giving too great a proportion of maize, either as meal or corn, or the effect will be a useless and prejudicial fattening from the large quantity of oil it contains; it is best mixed with barley or bean-meal, and is then a most economical and useful food. Potatoes, also, from the large proportion of starch contained in them, are not good unmixed as a regular diet for poultry; but mixed with bran or meal will be found most conducive to condition and laying.

In mixing soft food, there is one general rule always to be observed: it must be mixed rather *dry*, so that it will break if thrown upon the ground. There should never be enough water to cause the food to glisten in the light, or to make a sticky porridgy mass, which clings round the beaks of the fowls and gives them infinite annoyance, besides often causing diarrhœa.

If the weather be dry, and the birds are fed in a hard gravelled yard, the food is just as well, or better, thrown on the ground.

Fig. 3.

If they are fed in the shed, however, it is best to use an oblong dish of zinc, or, preferably, earthenware, such as represented in Fig. 3. The trough or dish must, however, be protected, or the fowls will walk upon it, scratch earth into it, and waste a large portion;

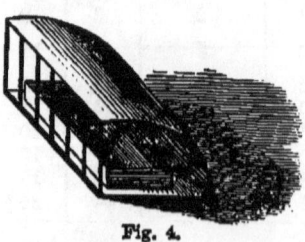

Fig. 4.

and this is best prevented by having a loose curved cover made of tin and wire, as shown in Fig. 4, which, when placed on the ground over the dish, will effectually prevent the fowls having anything to do with the food except to eat it, which they are quite at liberty to do through the perpendicular wires, two and a-half inches apart. Many experienced poultry-keepers prefer to drive the wires into the ground, leaving them six inches high; the trough is then put behind them, and a board laid over, leaning on the top of the wires. The effect of such a plan is precisely similar as regards the protection of the food, and its only disadvantage is, that the wires being always in the ground rather hinder the sweeping of the shed. For this reason we contrived the above cover, and consider it the best, as it is certainly the most convenient plan.

If the fowls have a field to run in they will require no further feeding till their evening meal of grain. Taking it altogether, no grain is more useful or economical than barley, and in summer this may be occasionally changed with oats; in winter, for the reasons already given, Indian corn may be given every second or third day with advantage. Buckwheat is, chemically, almost identical in composition with barley, but it certainly has a stimulating effect on the production of eggs, and it is a pity it cannot be more frequently obtained at a cheap rate. We never omit purchasing a sack of this grain when we *can*, and have a strong opinion that the enormous production of eggs and fowls in France is to some extent connected with the almost universal use of buckwheat by French poultry-keepers. Wheat is generally too dear to be employed, unless damaged, and if the damage be great it had better not be meddled with; but if only slightly injured, or if a good sample be offered of light "tail" wheat, as it is called, it is a most valuable food, both for chickens and fowls. "Sweepings" sometimes contain poisonous substances; are invariably dearer, weight for weight, than sound grain; and should never be seen in a poultry-yard.

The mid-day meal of penned-up fowls should be only a scanty one, and may consist either of soft food or grain, as most convenient—meal preferably in cold weather.

The regular and substantial diet is now provided for, but will not alone keep the fowls in good health and laying. They are omnivorous in their natural state, and require some portion of *animal* food. On a wide range they will provide this for themselves, and in such an establishment as figured at page 11, the scraps of the dinner-table will be quite sufficient; but if the number kept be large, with only limited accommodation, it will be necessary to buy every week a few pennyworth of bullocks' liver, which may be boiled, chopped fine, and mixed in their food, the broth being used instead of water in mixing; these little tit-bits will be eagerly picked out and enjoyed. A very little is all that is necessary, and need not be given more than three times a week. If fowls be much over-fed with this kind of food the quills of the feathers become more or less charged with blood, which the birds in time perceive, and almost invariably peck at each other's plumage till they leave the skin quite bare. It is also necessary to give a caution against the use of greaves, so much recommended, for obvious reasons, by the vendors. When fowls are habitually fed upon this article their feathers speedily become disarranged and fall off, and when killed the flavour, to any ordinary palate, is disgusting.

There is yet another most important article of diet, without which it is absolutely *impossible* to keep fowls in health. We refer to an ample and daily supply of green or fresh vegetable food. It is not perhaps too much to say, that the omission of this is the proximate cause of nearly half the deaths where fowls are kept in confinement; whilst with it, our other directions having been observed, they may be kept in health for a long time in a pen only a few feet square. It was to provide this that we recommended the open yards, in page 11, to be laid down in grass—the very best green food for poultry; and a run of even an hour daily on such a grass plot, supposing the shed to be dry and clean, will keep them in vigorous health, and not be more than the grass will bear. But if a shed only be available, fresh vegetables must be thrown in daily. Anything will do. A good plan is to mince up cabbage-leaves or other refuse vegetables, and mix pretty freely with the soft food; or the whole leaves may be thrown down for the fowls to devour; or a few turnips may be minced up daily, and scattered like grain, or simply cut in two and thrown into the run; or if it can be got, a large sod of fresh-cut turf thrown to the fowls will be better than all. But something they *must* have every day, or nearly so, otherwise their bowels sooner or later become disordered, their feathers look dirty, and their combs lose that beautiful bright red colour which will always accompany really good health and condition, and testifies pleasantly to abundance of eggs.

The water vessel must be filled fresh every day at least, and so arranged that the birds cannot scratch dirt into it, or make it foul. The ordinary poultry-fountain is too well known to need description, but a rather better form than is usually made is shown in Fig. 5. The advantages of such a construction are two: the top being open, and fitted with a cork, the state of the interior can be examined, and the vessel well sluiced through to remove the green slime which always collects by degrees, and is very prejudicial to health; and the trough being slightly raised from the ground, instead of upon it, the water is less easily fouled. But either form, if placed with the trough *towards* the wall, at a few inches distance from it, will keep the water clean very well. Some experienced breeders prefer shallow pans; but if these be adopted they must be either put behind rails, with a board over, or protected by a cover, in the same way as the feeding troughs already described.

Fig. 5.

Fowls must *never* be left without water. During a frost, therefore, the fountain should be emptied every night, or there will be trouble next morning. Care must always be taken also that *snow* is not allowed to fall into the drinking vessel. The reason has puzzled wiser heads than ours; but it is a *fact*, that any real quantity of snow-water seems to reduce both fowls and birds to mere skeletons.

It is well in winter to add to the water a few drops of a solution of sulphate of iron (green vitriol), just enough to give a slight mineral taste. This will, in a great measure, guard against roup, and act as a bracing tonic generally. The *rusty* appearance the water will assume is quite immaterial. The best plan, perhaps, is to keep a large bottle of the celebrated "Douglas* mixture," respecting which we can speak with unqualified approval, as a most valuable addition to the drink in cold weather of both fowls and chickens. It consists of half a pound of sulphate of iron and one ounce of sulphuric acid dissolved in two gallons of water; and is to be added in the proportion of a tea-spoonful to each pint of water in the fountain.

Whilst the fowls are moulting, the above mixture, or a little sulphate of iron, should always be used; it will assist them greatly through this, the most critical period of the whole year. A little hemp-seed should also be given every day at this season, at least to all fowls of value; and with these aids, and a little pepper on their food, with perhaps a little *extra* meat, or even a little *ale* during the few weeks the process lasts, there will rarely be any lost. With hardy kinds and good shelter such precautions are scarcely necessary, but they cost little, and have their effect also on the early recommencement of laying.

In addition to their regular food it will be needful that the fowls have a supply of *lime*, in some shape or other, to form the shells of their eggs. Old mortar pounded is excellent; so are oyster-shells well burnt in the fire and pulverised; of the latter they are very fond, and it is an excellent plan to keep

* So called because published in the *Field* newspaper by Mr. John Douglas, then superintending the Wolseley Aviaries.

a "tree-saucer" full of it in their yard. If this matter has been neglected, and soft shell-less eggs have resulted, the quickest way of getting matters right again is to add a little lime to the drinking water.

We shall conclude this chapter with a few further remarks respecting general management.

With regard to the nests, they may be of any form, but are best upon the ground. A long box may be employed, divided by partitions into separate compartments; or separate laying-boxes may be used, which is preferable, as more easily cleaned. Many like baskets, made flat on one side, and hung to a nail in the wall; these should be of wire, and then cannot harbour vermin—the great plague of fowls. The straw should be broken and beaten till it is quite soft, and changed as often as there is any foul or musty smell. If the nests are offensive the hens will often drop their eggs, quite perfect, upon the ground rather than resort to them.

Cleanliness in the house and run has already been insisted upon, and is only again alluded to on account of the value of the manure. This, collected daily, should be put in any convenient receptacle where it can be kept dry, and either used in the garden, if there is one, or sold. It pays best to use it where possible; it should always be mixed with earth, being very strong, and is especially valuable for all plants of the cabbage kind; it is also excellent for growing strawberries, or indeed almost anything if sufficiently diluted. If there be no possibility of so using it, it is worth at least seven shillings per cwt. to sell, and is greatly valued by all nurserymen and gardeners who know its value; but there is sometimes difficulty in finding those who do, and getting a fair price. The lowest price we ever knew offered, however, was three shillings per cwt. At seven shillings (which we believe to be about a fair value, compared with that of guano, on account of the moisture contained) we consider the value of the manure equal to fully one-fifth—perhaps one-fourth would be nearer the mark—of the total profit from the fowls. It is, therefore, an item too important to be neglected.

Where a considerable number of fowls are killed annually the feathers also become of value, and should be preserved. They are very easily dressed at home. Strip the plumage from the quills of the larger feathers, and mix with the small ones, putting the whole loosely in paper bags, which should be hung up in the kitchen, or some other warm place, for a few days to dry. Then let the bags be baked three or four times, for half an hour each time, in a cool oven, drying for two days between each baking, and the process will be completed. Less trouble than this will do, and is often made to suffice; but the feathers are inferior in crispness to those so treated, and may occasionally become offensive.

Eggs should be collected regularly, if possible twice every day; and if any chickens are to be reared from the home stock, the owner or attendant should learn to recognise the egg of each particular hen. There is no difficulty in this, even with a considerable number—nearly every egg, to the accustomed eye, has a well-marked individual character; and if there be any hens of value, it may save much disappointment in the character of the brood to know the parentage of those selected for hatching.

Before concluding, it may be expected that something definite should be said respecting the actual profit of what may be called *domestic* poultry-keeping. It is extremely difficult to make any such statement, so much depends upon the price of food, upon the management, selection of stock, and value of eggs. But in general we have found the average cost of fowls, when properly fed, to be about 1d. per week each for ordinary sorts, and not exceeding 1½d. per week for the larger breeds; when the cost is more we should suspect waste. A good *ordinary* hen ought to lay 120 eggs in a year. and if good laying

IMPORTANCE OF SYSTEM.

breeds are selected, such as we have named in Chapter II., there ought to be an average of fully 150, not reckoning the cock, whilst Game or Hamburghs will exceed 200 per annum. Of course, good management is supposed, and a regular renewal of *young* stock, as already insisted upon. For domestic purposes eggs ought to be valued at the price of new-laid, and from these data each can make his own calculation. The value of the manure, when it can be sold or used, we consider is about 9d. to 1s. per annum for each fowl.

Finally, let the whole undertaking—large or small—be conducted as a real matter of business. If more than three or four hens are kept, buy the food wholesale, and in the best market; let the grain be purchased a sack at a time—potatoes by the cart-load or hundred-weight, and so on. Let a fair and strict *account* be kept of the whole concern. The scraps of the house may be thrown in, and the cost of the original stock, and of their habitations, may be kept separate, and reckoned as capital invested; but let everything afterwards for which *cash* is paid be rigorously set down, and on the other side, with equal strictness, let every egg or chicken eaten or sold be also valued and recorded. This is of great importance. The young beginner may, perhaps, manage his laying-stock well, but succeed badly with his chickens (though not, we hope, if he be a reader of this book), or *vice versa ;* and it is no small matter in poultry-keeping, as in any other mercantile concern, to be able to see from recorded facts *where* has been the profit or where the loss. The discovery will lead to reflection; and the waste, neglect, or other defective management being amended, the hitherto faulty department will also contribute its quota to the general weal.

CHAPTER IV.

INCUBATION.

Much disappointment in the hatching and rearing of young broods would be prevented were more care taken that the eggs selected for setting were of good quality—not only likely to be fertile, but the produce of strong and hardy birds. This remark applies to common barn-door poultry quite as much as to the pure breeds. A friend recently complained to us, that out of a dozen eggs only four or five had hatched; and on inquiry, we found that the sitting had been procured from an inn-yard, where, to our own knowledge, only one cock was running with about twenty hens, from which of course no better result could be expected. When the eggs have to be procured from elsewhere, therefore, whatever be the class of fowls required, it should first of all be ascertained that there is at least one cock to every six or eight hens, and that he be a strong and lively bird; and next, that the fowls be not only of the kind desired, but that they are well fed and taken care of. From scraggy, half-starved birds it is impossible to rear a large brood, as the greater number even of those hatched will die in infancy. It only remains to ensure that the eggs be *fresh*, and a successful hatching may be anticipated.

With regard to this latter point, eggs have been known to hatch when two months old, or even more; but we would never ourselves set, from choice, any egg which had been laid more than a fortnight; and after a month, or less, it is useless trouble. Fresh eggs, if all be well, hatch out in good time, and the chicks are strong and lively; the stale ones always hatch last, being perhaps as much as two days later than new-laid, and the chickens are often too weak to break the shell. We have also invariably noticed, when compelled to take a portion of stale eggs to make up a sitting, that even when such eggs have hatched, the subsequent deaths have principally occurred in this portion of the brood; but that if none of the eggs were more than four or five days old, they not only hatched nearly every one, and within an hour or two of each other, but the losses in any ordinary season were very few.

When the eggs are from the home stock, their quality should, of course, be above suspicion. It is scarcely necessary to say, that in order to ensure this, every egg before storing should have legibly written upon it in pencil the date on which it was laid. Eggs intended for setting are best kept in bran, the large end downward, and should never be exposed to concussion. Another very good plan is to have a large board pierced with a number of round holes in regular rows to receive the eggs.

Hundreds of years ago it was thought that the sex of eggs could be distinguished by the shape—the cocks being produced from those of elongated shape, and hens from the short or round. Others have pretended to discern the future sex from the position of the air-bubble at the large end. We need scarcely say, that these and every other nostrum have, hundreds of times, been proved to be erroneous. There is not a breeder of prize poultry in England who would not gladly give twenty pounds for the coveted knowledge, and thenceforth breed no more cockerels than he really wanted; but the secret has never been discovered yet, and it is even impossible to tell before the egg has been sat upon a short time whether it has been fecundated.

We have, in a previous chapter, already mentioned that the sitting hens *ought* to have a separate shed and run provided for them, in order that the other hens may not occupy their nests during absence, or they themselves go back to the wrong ones, as they will often do if allowed to sit in the fowl-house. Even in a very small domestic establishment we strongly recommend that the small additional space requisite be devoted to this purpose, for all our experience has proved that, whatever success may be obtained otherwise by constant care and watchfulness, it is never so great as when the sitter can be shut into a separate run, and be *entirely unmolested*. An extensive run is neither necessary nor desirable, as it only entices the birds to wander, whereas, in a limited space, they will go back to their nests as soon as their wants are satisfied. A shed five feet square, with a run the same width for ten feet out in front, is quite sufficient for three hens.

If the hen *must* be set on the ordinary nest in the fowl-house, unless she can be watched every day to see that all goes right, it is best to take her off at a regular time every morning, and after seeing to her wants and due return, to shut her in so that she cannot be annoyed. She should be lifted by taking hold under the wings, gently raising them first to see that no eggs are enclosed. Very fair success may be attained by this method of management, which is obviously almost imperative in very large establishments, where numerous hens must be sitting at one time; but where such large numbers do not allow of a special poultry attendant it is rather troublesome, and on an average there will be a chicken or two less than if the hens can be put quite apart, where they need neither be watched nor interfered with. Since we adopted this plan we have, from good eggs, always hatched at least nine out of twelve, and generally more; and have had no trouble nor anxiety till the broods were actually hatched, which is anything but the case on the other system.

With respect to the arrangement of the hatching run, it should, if possible, be in sight of the other fowls, as it will keep the sitter from becoming strange to her companions, and prevent an otherwise inevitable fight on her restoration, to the possible damage of the brood. We prefer ourselves, as stated in the first chapter, a shed five feet wide and five deep, *open* in front to a small gravel or grass run. Under the shed must be, besides the nests, a good-sized shallow box of sand, dry earth or fine coal ashes, for the hen to cleanse herself in, which she specially needs at this time; and food and water must be *always* ready for her. With these precautions the hen may and should in nearly every case, with the exceptions presently mentioned, be left entirely to herself. There are, however, some birds which, if not removed, would starve upon their nests sooner than leave them; and therefore if the hen has not been off for two or three days (we would test her for that time first), we should certainly remove the poor thing for her own preservation. To feed upon the nest is a cruel practice, which has crippled many a fowl for life, and cannot be too strongly condemned.

Of all mothers we prefer Cochins or Brahmas. Their abundant "fluff" and feathering is of inestimable advantage to the young chicks, and their tame and gentle disposition makes them submit to any amount of handling or management with great docility. Cochins certainly appear clumsy with their feet, but we have never found more chickens actually trodden upon by them than with any other breed. Many complain that they leave their chickens too soon, but we have not found it so ourselves. If they are kept cooped instead of being set at liberty they will brood their chickens for at least two months, even till they have laid a second batch of eggs and desire to sit again; and by that time any brood is able to do without a mother's care. With regard to Brahmas as mothers, they have a peculiarity we never observed in any other fowl, and have never seen noticed in any work on poultry —they actually appear to *look behind them* when moving, lest they should tread upon their little ones. Dorkings, also, are exemplary mothers, and go with their chickens a long time, which recommends them strongly for very early broods. And lastly, a Game hen has qualities which often make her most valuable. She is not only exemplary in her care, and a superexcellent forager for her young brood, but will defend them to the last gasp, and render a good account of the most determined cat that ever existed; indeed, we would almost defy any single creature whatever, quadruped or otherwise, to steal a chick in daylight from a well-bred Game hen. But whatever be the hen chosen, she should be well feathered, moderately short-legged, and tolerably tame. Very high authority* has affirmed that only mature hens should be allowed to sit, and that pullets are not to be trusted; but our own experience and that of very many large breeders does not confirm this. We have constantly set pullets, and never had any more reason to complain of them than of older birds.

The nests may be arranged under the shed any way so that no one can see into them, with the one proviso that they be actually *upon the ground*. Chicks thus obtained always show more constitution than those hatched on a wooden bottom at a higher level. This holds good even at all times of the year. We are aware that eminent authorities who recommend ground-nests in summer, prefer a warm, wooden box in winter for the sake of the hen; but she will rarely suffer. The heat of her body whilst sitting is so great that a cool situation seems grateful to her—at least, a hen set on the ground rarely forsakes her nest, which is otherwise no uncommon case. We knew of a hen which, during the month of January, made her nest upon the top of a rock in one of the highest and most exposed situations in the Peak of Derbyshire, and brought a large brood of strong chickens into the yard. It is only necessary the birds should be protected from wind and rain, in order to avoid rheumatism; and this is most effectually done by employing for the nest a tight wooden box, like Fig. 6, open at

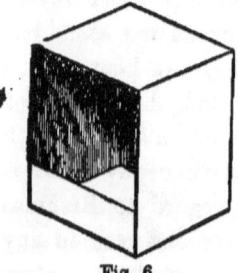

Fig. 6.

* Mrs. Fergusson Blair.

IMPORTANCE OF MOISTURE.

the bottom, and also in front, with the exception of a strip three inches high to contain the straw. Let one of these boxes be placed in the back corner of the shed, touching the side, the front being turned to the back wall, and about nine inches from it; and the hen will be in the strictest privacy, will be both perfectly sheltered and kept cool, and will never mistake her own nest for the one which may be placed in the other corner. If a third must be made room for, let her nest be placed the same distance from the wall midway between the others, and like them, with the front of the nest to the back of the shed. There will then be still nearly a foot between each two nests for the birds to pass.

A damp situation is best for the sitting shed, and will ensure good hatching in hot weather, when perhaps all the neighbours are complaining that their chicks are dead in the shells. Attempting to keep the nest and eggs *dry* has ruined many a brood. It is not so in nature; every morning the hen leaves her nest, and has to seek her precarious meal through the long, wet grass, which drenches her as if she had been ducked in a pond. With this saturated breast she returns, and the eggs are duly moistened. But if the nest be dry, the hen be kept dry, and the weather happen to be hot and dry also, the moisture within the egg itself becomes dried to the consistency of glue, and the poor little chick, being unable to *move round* within the shell, cannot fracture it, and perishes. Such a mishap will not happen if the ground under the nest be damp and cool. All that is necessary in such a case is to scrape a slight hollow in the bare earth, place the nest-box, already described, over it, and put in a moderate quantity of straw cut into two-inch lengths; or, still better, some fresh-cut damp grass may be put in first, and the straw over. Shape the straw also into a very *slight* hollow, and the nest is made; but care must be taken to well fill up the corners of the box, or the eggs may be rolled into them and get addled. Some prefer to put in first a fresh turf; but if the nest be on the bare ground, as we recommend, this is useless. In any case, the straw should be cut into short lengths for a hatching nest, and the neglect of this precaution is the most frequent cause of breakage; the hen, during her twenty-four hours' stay, gets her claws entangled in the long straws, and on leaving for her daily meal is very likely to drag one or two with her, fracturing one or more eggs, or even jerking them quite out of the nest.

Should such a mishap occur (and the nest should be examined every two or three days, when the hen is absent, to ascertain), the eggs must be removed, and clean straw substituted, and every sound egg at all soiled by the broken one be washed with a sponge and warm water, gently but quickly drying after with a cloth. The hen, if very dirty, should also have her breast cleansed, and the whole replaced *immediately*, that the eggs may not be chilled. A moderate hatch may still be expected, though the number of chicks is always more or less reduced by an accident of this kind. If, however, the cleansing be neglected for more than a couple of days after a breakage, or less at the latter period of incubation, probably not a single chick will be obtained; whether from the pores of the shell being stopped by the viscid matter, or from the noxious smell of the putrefying egg, it is not very material to inquire.

Every egg should also be marked quite round with ink or pencil, so that if any be subsequently laid in the nest they may be at once detected and removed. Hens will sometimes lay several eggs after beginning to sit.

In ordinary winters the hen should be set as in summer, giving her, however, rather more straw. Only in severe frost should she be brought into the house; and in that case, or in summer if the ground be very dry, it will be necessary during the last half of the hatching period to sprinkle the eggs slightly with water every day while she is off. This is done best by

TESTING THE EGGS.

dipping a small brush in tepid water; and is *always* necessary to success, in dry weather at least, when a hen is set in a box at a distance from the ground, as is the case in large sitting houses. But, where it can be had, we much prefer the natural moisture of a damp soil: it never fails, and avoids going near the hen.

When the number of eggs set yearly is considerable, it is worth while to withdraw the unfertile ones at an early period. About the eighth day let the hen be removed by candlelight, and each egg be held between the eye and the light, in the manner represented by Fig. 7. If the egg be fertile, it will appear opaque, or dark all over, except, perhaps, a small portion towards the top; but if it be unimpregnated, it will be still translucent, the light passing through it almost as if new laid. After some experience the eggs can be distinguished at an earlier period, and a practised hand can tell the unfertile eggs even at the fourth day. Should the number withdrawn be considerable, four batches set the same day may be given to three hens, or even two, and the remainder given fresh eggs; and if not, the fertile eggs will get more heat, and the brood come out all the stronger. The sterile eggs are also worth saving, as they are *quite good* enough for cooking

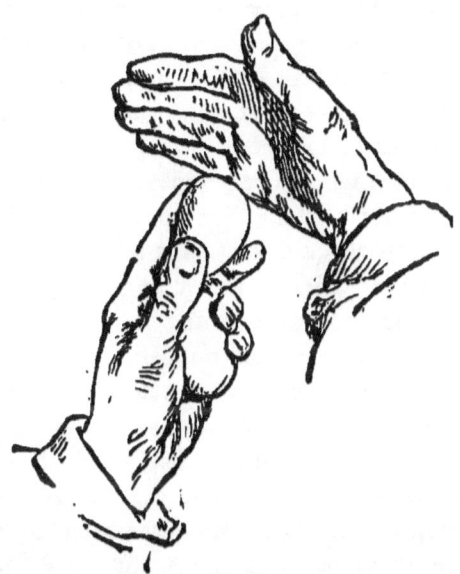

Fig. 7.

purposes, and quite as fresh even for boiling as nine-tenths of the Irish eggs constantly used for that purpose. We do not, however, recommend this plan when the sitters are few and the eggs from the home stock, as in that case their quality should be known, and sterility very rare.

It is a common mistake to set too many eggs. In summer, a large hen may have thirteen, or a Cochin fifteen of *her own;* but in early spring eleven are quite enough. We have not only to consider how many chickens the hen can hatch, but how many she can *cover* when they are partly grown. If a hen be set in January, she should not have more than seven or eight eggs, or the poor little things, as soon as they begin to get large, will have no shelter, and soon die off. It is far better to hatch only six and rear five, or may be all, to health and vigour, than to hatch ten and only probably rear three puny little creatures, good for nothing but to make broth. In April and May broods, such a limitation is not needed; but even then eleven or twelve chickens are quite as many as a large, well-feathered hen can properly nourish, and the eggs should only be one or two in excess of that number.

A good hen will not remain more than half an hour away from her nest, unless she has been deprived of a dust-bath, and so become infested with lice, which sometimes causes hens thus neglected to forsake their eggs altogether. When a hen at the proper time shows no disposition to return, she should be quietly driven towards her nest; if she be caught, and replaced by hand, she is often so frightened and excited as to break the eggs. A longer absence is not, however, necessarily fatal to the brood. We have had hens repeatedly absent more than an hour, which still hatched seven or eight chicks; and on one occasion a hen sitting in the fowl-house returned to the wrong nest, and was absent from her own more than five hours. We of course considered all chances of hatching at an end; but as the hen had been sitting a fortnight, concluded to let her finish her time, and she hatched five chickens. We have heard of a few hatching even after *nine* hours' absence, and therefore would never, on account of such an occurrence, abandon valuable eggs without a trial.

The chickens break the shell at the end of the twenty-first day, on an average; but if the eggs are new-laid, it will often lessen the time by as much as five or six hours, while stale eggs are always more or less behind.

We never ourselves now attempt to assist a chick from the shell. If the eggs were fresh, and proper care has been taken to preserve moisture during incubation, no assistance is ever needed. To fuss about the nest frets the hen exceedingly; and we have always found that even where the poor little creature survived at the time, it never lived to maturity. Should the reader attempt such assistance, in cases where an egg has been long "chipped," and no further progress made, let the *shell* be cracked gently all round, without tearing the inside membrane; if *that* be perforated, the viscid fluid inside dries, and glues the chick to the shell. Should this happen, or should both shell and membrane be perforated at first, introduce the point of a pair of scissors, and cut *up* the egg towards the large end, where there will be an empty space, remembering that if blood flow all hope is at an end. Then put the chick back *under the hen;* she will probably squeeze it to death, it is true, it is so very weak; but it will *never* live if put by the fire, at least, we always found it so. Indeed, as we have said, we consider it quite useless to make the attempt at all.

But with good eggs, a good hen, and good management, all will go right, and there will be in due time a goodly number of strong and healthy chickens, to the mutual delight of the hen and of her owner. And with the treatment of the young brood we will begin another chapter.

CHAPTER V.

THE REARING AND FATTENING OF CHICKENS.

For nearly twenty-four hours after hatching, chickens require no food at all; and though we do not think it *best* to leave them quite so long as this without it, we should let them remain for at least twelve hours undisturbed. We say undisturbed, because it is a very common practice to take those first hatched away from the hen, and put them in a basket by the fire till the whole brood is out. When the eggs have varied much in age, this course *must* be adopted; for some chickens will be perhaps a whole day or more behind the others, and the hen, if she felt the little things moving beneath her, would not stay long enough to hatch the rest. But we have explained in the last chapter that this should not be, and that if the eggs are all fresh, the chicks will all appear within a few hours of each other. In that case they are much better *left with their mother:* the heat of her body appears to strengthen and nourish them in a far better manner than any other warmth, and they are happy and contented, instead of moving restlessly about as they always do whilst away from her.

Our own plan is to set the eggs in the evening, when the chicks will break the shell in the evening also, or perhaps the afternoon. Then at night let the state of the brood be *once* only examined, all egg-shells removed from the nest, and the hen, if she be tame enough to receive it, given food and water. Let her afterwards be so shut in that she cannot leave her nest, and all may be left safely till the morning. By that time the chicks will be strong and lively, quite ready for their first meal; and unless some of the eggs are known to be very stale, any not hatched then are little likely to hatch at all. If this be so, the chicks may be removed and put in flannel by the fire, and another day patiently waited, to see if any more will appear. We should not do so, however, if a fair number had hatched well; for they never thrive so well away from the hen, and it is scarcely worth while to injure the healthy portion of the brood for the sake of one or two which very probably may not live after all.

The first meal should be given *on the nest*, and the best material for it is an equal mixture of hard-boiled *yolk* of egg and stale bread-crumbs, the latter slightly moistened with milk. Let the hen be allowed to partake of this also—she needs it; and then give her besides as much barley as she will eat, and offer her water, which she will drink greedily. To satisfy the hen *at first* saves much restlessness and trouble with her afterwards.

There is a stupid practice adopted by many, of removing the little horny scale which appears on every chicken's beak, with the idea of enabling them to peck better, and then to put food or pepper-corns down their throats, and dip their bills in water to make them drink. It is a mistake to say that if this does no good it can do no harm : the little beaks are very soft and tender, and are often injured by such barbarous treatment. *Leave them alone.* If they do not eat or drink—and chickens seldom drink the first day—it only shows they do not wish to ; for to fill an empty stomach is the first and universal instinct of all living things.

The brood having been fed, the next step will depend upon circumstances. If, as we recommend, the chickens were hatched the night before, or be well upon their legs, and the weather be fine, they may be at once moved out, and the hen cooped where her little ones can get the sun. If it be winter, or settled wet weather, the hen must, if possible, be kept on her nest this day also, and when removed be cooped in a dry shed or outhouse.

The best arrangement, where there is convenience for it, is

that shown in Fig. 8. A shed, six feet square, is reared against the wall, with a southern exposure, and the coop placed under it. This coop should be made on a plan very common in some parts of France, and consists of two compartments, separated by a partition of bars; one compartment being closed

Fig. 8.

in front, the other fronted with bars like the partition. Each set of bars should have a sliding one to serve as a door, and the whole coop should be tight and sound. It is best to have no bottom, but to put it on loose dry earth or ashes, an inch or two deep. Each half of the coop is about two feet six inches square, and may or may not be lighted from the top by a small pane of glass.

The advantage of such a coop and shed is, that except in very severe weather, no further shelter is required even at night. During the day the hen is kept in the outer compartment, the chickens having liberty, and the food and water being placed outside; whilst at night she is put in the inner portion of the coop, and a piece of canvas or sacking hung over the bars of the outer half. If the top be glazed, a little food and the water vessel may be placed in the outer compartment at night, and the chicks will be able to run out and feed early in the morning, being prevented by the canvas from going out into the cold air. It will be only needful to remove the coop every two days for a few minutes, to take away the tainted earth and replace it with fresh. There should, if possible, be a grass-plot in front of the shed, the floor of which should be covered with dry loose dust or earth.

Under such a shed chickens will thrive well; but if such cannot be obtained, sufficient shelter during ordinary breeding seasons may be obtained by the use of a well-made board coop, with a gabled roof covered with felt. This coop should be open in front only, and be two feet six inches or three feet square. At night let a thick canvas wrappering be hung over the front.

The ordinary basket coop is only fit to be used in perfectly fine weather, when it is convenient to place on a lawn. Some straw, weighted by a stone, or other covering, should, however, be placed on the top, to give shelter from the mid-day sun.

It is often necessary in considerable establishments to carry the hen and her brood for a considerable distance. For this purpose the box shown in Fig. 9 will be found very convenient. It may be made in either one or two divisions, and the chickens will be thus managed with no trouble, as they cannot escape when put in at the top, whilst they are readily let out again by the door.

Chickens should always, *if possible*, be cooped near grass. No single circumstance is so conducive to health, size, and vigour, supposing them to be decently well cared for, as even a small grass run such as that provided in Fig. 2. Absolute cleanliness is also essential, even more than for grown fowls; and the reason why difficulty is often experienced in rearing large numbers is, that the ground becomes so tainted with their excrements. The coop should, therefore, be either moved to a fresh place every day, or the dry earth under be carefully

Fig. 9.

removed. A very good plan, and one we have found in a limited space to answer remarkably well, is to have a wooden gable-roofed coop made with a wooden bottom, and to cover this an *inch deep* with perfectly dry earth, or fine sifted ashes. The ashes are renewed every evening in five minutes, and form a nice warm bed for the chicks, clean and sweet, and much better than straw.

Cats sometimes make sad inroads on the broods. If this nuisance be great, it is well to confine the coveted prey while young within a wire-covered run. And the best way of forming such a run, is to stretch some inch-mesh wire-netting, two feet wide, upon a light wooden frame, so as to form two wire hurdles, two feet wide and about six feet long, with one three feet long. These are easily lashed together with string to form a run six feet by three (Fig. 10), and may be covered by a similar hurdle of two-inch mesh three feet wide. In such a run all animal depredations may be defied; and in any case we should recommend its use until the chicks are a fortnight old; it saves a world of trouble and anxiety, and prevents the brood wandering and getting over-tired. By having an assortment of such hurdles, portable runs can be constructed in a few minutes of any extent required, and will be found of great

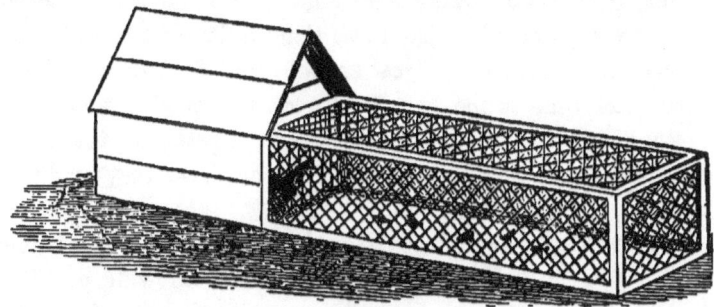

Fig. 10.

advantage until the broods are strong. The hen may also be given her liberty within the prescribed bounds.

With regard to feeding, if the question be asked what is the *best* food for chickens, irrespective of price, the answer must decidedly be oatmeal. After the first meal of bread-crumbs and egg no food is equal to it, if *coarsely* ground, and only moistened so much as to remain crumbly. The price of oatmeal is, however, so high as to forbid its use in general, except for valuable broods; but we should still advise it for the first week, in order to lay a good foundation. It may be moistened either with water or milk, but in the latter case only sufficient must be mixed for each feeding, as it will turn sour within an hour in the sun, and in that condition is very injurious to the chickens.

For the first three or four days the yolk of an egg boiled hard should also be chopped up small, and daily given to each dozen chicks; and when this is discontinued, a little cooked meat, minced fine, should be given once a day till about three

E

weeks old. The cost of this will be inappreciable, as a piece the size of a good walnut is sufficient for a whole brood, and the chickens will have more constitution and fledge better than if no animal food is supplied.

Food must be given very often. For the first week every hour is not too much, though less will do; the next three weeks, every two hours; from one to two months old, every three hours; and after that, three times a day will be sufficient. To feed *very often*, giving just enough *fresh* food to be entirely eaten each time, is the one great secret of getting fine birds. If the meals are fewer, and food be left, it gets sour, the chicks do not like it, and will not take so much as they ought to have.

After the first week the oatmeal can be changed for cheaper food. We can well recommend any of the following, and it is best to change from one to another, say about every fortnight. An equal mixture of "sharps" and barley-meal, or "sharps" and buckwheat-meal, or of bran and Indian meal; or of bran, oatmeal, and Indian meal. The last our own chickens like best of all, and as the cheap bran balances the oatmeal, it is not a dear food, and the chicks will grow upon it rapidly. Potatoes mashed with bran are also most excellent food.

The above will form the staple food, but after a day or two some grain should be given in addition. Groats chopped up with a knife are excellent; so is crushed wheat or bruised oats. Chickens seem to prefer grits to anything, but it is not equal to meal as a permanent diet. A little of either one or the other should, however, be given once or twice a day, and in particular should form the last meal at night, for the reasons given in page 23.

Bread sopped in water is the worst possible food for chickens, causing weakness and general diarrhœa. With milk it is better, but not equal to meal.

Green food is even more necessary to chickens than to

adult fowls. Whilst very young it is best to cut some grass into very small morsels for them with a pair of scissors; afterwards they will crop it for themselves if allowed. Should there be no grass plot available, cabbage or lettuce-leaves must be regularly given—minced small at first, but thrown down whole as soon as the beaks of the chickens are strong enough to enable them to help themselves.

In winter or very early spring the chickens must, in addition to the above feeding, have more stimulating diet. Some under-done meat or egg should be continued regularly, and it is generally necessary to give also, once a day at least, some stale bread soaked in ale. They should also be fed about eight or nine o'clock, by candle-light, and early in the morning. In no other way can Dorkings or Spanish be successfully reared at this inclement season, though the hardier breeds will often get along very well with the ordinary feeding. Ale and meat, with liberal feeding otherwise, will rear chickens at the coldest seasons; and the extra cost is more than met by the extra prices then obtained in the market. But shelter they *must* have; and those who have not at command a large outhouse or shed to keep them in while tender, should not attempt to raise winter or early spring chickens—if they do, the result will only be disappointment and loss. The broods should only be let out on the open gravel or grass in bright, or at least *clear*, dry weather.

At the age of four months the chickens, if of the larger breeds, should be grown enough for the table; and if they have been well fed, and come of good stock, they will be. For ourselves we say, let them be eaten as they are—they will be quite fat enough; and fattening is a very delicate process, success in which it takes some experience to acquire. For market, however, a fatted fowl is more valuable; and the birds should be penned up for a further fortnight or three weeks, which ought to add at least two pounds to their weight. For a limited number of chickens it will be sufficient to provide a small number of simply constructed pens, such as are represented in Fig. 11. Each compartment should measure about nine by eighteen inches, by about eighteen inches high; and the bottom should not consist of board, but be formed of bars two inches wide placed two inches apart, the top corners being rounded off. The partitions, top and back, are board, as the

Fattening Pens. Fig. 11.

birds should not see each other. These pens ought to be placed about two inches from the ground in a darkish, but not cold or draughty place, and a shallow tray be introduced underneath, filled with fresh dry earth every day, to catch the droppings. This is the best and least troublesome method of keeping the birds clean and in good health. As fast as each occupant of a pen is withdrawn for execution its pen should be whitewashed all over inside, and allowed to get perfectly dry before another is introduced. This will usually prevent much trouble from insect vermin; but if a bird appears restless from that cause, some powdered sulphur, rubbed well into the roots of the feathers, will give immediate relief.

In front of each compartment should be a ledge three inches

wide, on which to place the food and water-tins. The latter must be replenished once, the former three times a day; and after each meal the pens must be darkened for *half* the time until the next, by hanging a cloth over the front. This cloth is best tacked along at the top, when it can be conveniently hung over or folded back as required. The two hours' darkness ensures quiet and thorough digestion; but it is not desirable, as most do, to keep the birds thus the *whole* time till the next meal, as the chickens will have a much better appetite on the plan we recommend.

The best food for fattening is buckwheat-meal, when it can be obtained; and it is to the use of this grain the French owe, in a great measure, the splendid fowls they send to market. If it cannot be procured, the best substitute is an equal mixture of Indian and barley-meal. Each bird should have as much as it will eat at one time, but no food left to become sour : a little barley may, however, be scattered on the ledge. The meal may be mixed with skim-milk if available. A little minced green food should be given daily, to keep the bowels in proper order.

In three weeks the process ought to be completed. It must be borne in mind that *fat* only is added by thus penning a chicken; the lean or flesh must be made before, and unless the chicken has attained the proper standard in this respect, it is useless even to attempt to fatten it. Hence the importance of high feeding from the very shell. The secret of rearing chickens profitably is, to get them ready for the table at the *earliest possible period*, and not to let them live a *single day after*. Every such day is a dead loss, for they cannot be *kept* fat; once up to the mark, if not killed they get feverish and begin to waste away again. To make poultry profitable, even on a small scale, everything must go upon system; and that system is, to kill the chickens the very day they are ready for it.

If extra weight and fat is wanted, the birds may be

crammed during the last ten days of the fattening period, but not before. The meal is to be rolled up the thickness of a finger, and then cut into pellets an inch and a half long. Each morsel must be dipped in water before it is put into the bird's throat, when there will be no difficulty in swallowing. The quantity given can only be learnt by experience.

For home use, however, nothing can equal a chicken never fattened at all, but just taken out of the yard. If well fed there will be plenty of good *meat*, and the *fat* of a fowl is to most persons no particular delicacy. In any case, however, let the chicken be fasted twelve hours before it is killed.

There are various modes of killing—all of them very effectual in practised hands. One is to give the bird a very sharp blow with a small but heavy stick behind the neck, about the second joint from the head, which will, if properly done, sever the spine and cause death very speedily. Another is to clasp the bird's head in the hand and swing the body round by it—a process which also kills by parting the vertebræ. M. Soyer recommends that the joints be *pulled* apart, which can easily be effected by seizing the head in the right hand, placing the thumb just at the back of the skull, and giving a smart jerk of the hand, the other, of course, holding the neck of the fowl. And lastly, there is the knife, which we consider, after all, the most merciful plan, as it causes no more pain than that occasioned by the momentary operation itself. We do not advocate cutting the throat; but having first hung up the bird by the legs, thrust a long, narrow, and sharp-pointed knife, like a long penknife, which is made for the purpose, through the back part of the roof of the mouth up into the brain. Death will be almost instantaneous, which is too seldom the case when dislocation is employed.

Fowls are easiest plucked at once, whilst still warm, and should be afterwards scalded by dipping them for just *one instant* in boiling water. This process will make any decent fowl look plump and nice, and poor ones, of course, ought not to be killed at all. They should not be "drawn" until the day they are wanted, as they will keep much longer without.

With respect to old fowls, in the market they are an abomination; but at home it is sometimes needful to use them. If so, let them be *boiled*. Unless very aged, they will then be tolerable eating; but if roasted, will be beyond most persons' power of mastication.

CHAPTER VI.

DISEASES OF POULTRY.

IF fowls are kept clean, and well sheltered from wind and wet; are not overfed, and have a due proportion of both soft and green food, with a never-failing supply of *clean* water, they will remain free from disease, unless infected by strangers. And when a fowl becomes ill, the best cure in nearly every case is *to kill it* before it is too bad to be eaten. Only in the case of valuable birds, which people are naturally unwilling to sacrifice, do we recommend much attempt at a cure, and even then only where the disease is so defined and evident that the treatment is sure. To prescribe for a fowl in the dark is one of the most hopeless speculations that can well be.

As this work is intended to be strictly practical, it is only for such well-defined complaints we shall prescribe; and in doing so, it is only justice to acknowledge the great services rendered in this matter to the whole poultry world by Mr. W. B. Tegetmeier. That gentleman has long made the diseases of fowls his peculiar study, and has been above all others successful in the treatment of them; and the greater part of this chapter is founded more or less directly upon his authority.

Besides actual diseases, there are certain *natural* ailments, as they may be called, to which all fowls may be subject, and which demand treatment.

Bad Fledging.—Chickens often droop and suffer much whilst their feathers are growing, especially in cold wet weather; and the breeds which feather most rapidly suffer most. This is probably one reason why Cochins and Brahmas, which fledge late and slowly, are so hardy. As soon as a brood appears drooping whilst the feathers grow, if it has not been done before, begin *at once* giving them a little meat every day, and some bread sopped in ale. A little burnt oyster-shell, pounded very fine, and added to their food, is also beneficial. Keep them out of the wet, above all things, and they will generally come round. This crisis seldom lasts more than a week or ten days; the chicks either die off, or recover their health and vigour.

Leg Weakness.—Highly-fed chickens which grow fast, bred from prize stock, are most subject to this; which simply arises from outgrowing their strength, and must be met accordingly by animal food and tonics. Give meat or worms every day, and unless it be cold weather, dip the legs for a few minutes daily in cold water. The prescription will be, three or four grains of ammonio-citrate of iron for each chicken, given every day, dissolved in the water with which the meal is mixed.

The above affection must not be confounded with *cramp* from cold and wet, which also makes the birds unable to walk, or even stand, but for which cold bathing would be most injurious. In this case, the only treatment is warmth, feeding meanwhile on meal mixed with ale, and always given warm. Under this regimen the bird will soon recover, unless the attack has been long unperceived and neglected.

Bad Moulting.—Old fowls sometimes suffer much at this season, especially if the precautions recommended in Chapter III. have been overlooked. These precautions contain the **only** effectual treatment. Give stimulating food, *warm*, every morning, and well peppered, with meat and ale every day, and keep under cover in wet weather. Add also iron, in the form of "Douglas Mixture," to the drinking water; and let some hemp-seed be given with the grain every evening. The birds, if not sunk too low, will then usually pull through. Fowls should not, however, be kept until old, except in the case of pets or valuable stock birds.

For actual diseases, it is well in all large establishments to have a weather-tight and well-ventilated house kept as a hospital, in which healthy fowls should *never be placed*. Roup, in particular, is so contagious, that even a recovered bird should be kept by itself for a few days before being restored to its companions.

Gapes is a fatal disease of chickens, and which we believe infectious; it is, at all events, epidemic. Unless perhaps thus communicated by others, it never occurs except there has been foul water, exposure to wet, and want of nourishing food. The disease consists—at least, so far as actual symptoms extend—in a number of small worms which infest the windpipe, and cause the poor chicken to gasp for breath. If taken early, it will be sufficient to give every day a morsel of camphor the size of a grain of wheat, and to put camphor in the drinking water; or a little turpentine may be given daily in meal; taking care, of course, that the deficiencies in diet and shelter be amended. In fully-developed cases, the worms must be removed by introducing a loop of horsehair into the trachea, and turning it round during withdrawal; the operation to be repeated several times, till all the worms appear to be extracted. A feather, stripped almost up to the top, may be used instead of the horsehair. The *frequent* occurrence of gapes is a disgrace to any poultry-yard.

Apoplexy occurs from over-feeding, and can seldom be treated in time to be of service. If the fowl, however, although insensible, do not appear actually dead, the wing may be lifted, and a large vein which will be seen underneath freely opened, after which hold the bird's head under a cold water tap for a few minutes. It is just possible it may recover; if so, feed sparingly on soft food only for a few days. In overfed hens, this disease usually occurs during the exertion of laying; if, therefore, a laying hen be found dead upon the nest, let the owner at once examine the remainder, and should they appear in too high condition, reduce their allowance of food accordingly.

Loss of Feathers is almost always caused either by want of green food, or having no dust-bath. Let these wants therefore be properly supplied, removing the fowls, if possible, to a grass run. For local application, Mr. Tegetmeier recommends mercurial ointment, but we ourselves prefer an unguent composed of sulphur and creosote. Nothing, however, will bring back the feathers before the next moult.

Roup is always caused by wet, or *very* cold winds. It begins with a common cold, and terminates in an offensive discharge from the nostrils and eyes, often hanging in froth about those organs. It is most *highly contagious*, the disease being, as we believe, communicated by the sickly fowl's beak contaminating the drinking water; therefore, let all fowls affected by it be at once put by themselves, and have a separate water-vessel. Keep them warm, and feed with meal only, mixed with hot ale instead of water; add "Douglas Mixture" to the water, and give daily, in a bolus of the meal, half a grain of cayenne pepper, with half a grain of powdered allspice, or one of Baily's roup pills. Give also half a cabbage-leaf every day, and wash the head and eyes morning and evening with very diluted vinegar, or a five-grain solution of sulphate of zinc. Mr. Tegetmeier's treatment is, to feed on oatmeal mixed with ale, and green food unlimited; washing the head with tepid water, and giving daily one grain sulphate of copper. We prefer the above. Roup runs its course rapidly, and in a week the bird will either be almost well, or so nearly dead that it had better be killed at once. It is *the* disease of poultry, and to be dreaded accordingly; fortunately, the symptoms are specific, and the treatment equally so.

Pip is no disease, and demands no treatment, being only analagous to "a foul tongue" in human beings. Cure the roup, or bad digestion, or whatever else be the real evil, and the thickening of the tongue will disappear too.

Diarrhœa may be caused either by cold, wet weather, with inadequate shelter; neglect in cleansing the house and run; or from the reaction after constipation caused by too little green food. Feed on warm *barley* meal; give *some* green food, but not very much; and at first administer, four times a day, three drops of camphorated spirit on a pill of meal. This will usually effect a cure. If the evacuations become coloured with blood, the diarrhœa has passed into *dysentery*, and recovery is almost hopeless. Mr. Tegetmeier's prescription is one grain each of opium and ipecacuanha, with five grains chalk; but the camphorated spirit is a better remedy.

Soft Eggs are generally caused by over-feeding the hens, and the remedy is then self-evident. It may, however, occur from want of lime, which must of course be supplied, the best form being calcined and pounded oyster-shells. Occasionally it is occasioned by fright, from being driven about, but in that case will right itself in a day or two. If *perfect* eggs are habitually dropped on the ground, the proprietor should see whether the nests do not need purifying. This leads us to

Insect Vermin, which can only be troublesome from gross neglect, either of the fowls or of their habitations. In the one case, the remedy is a dust-bath, mixed with powdered coke or sulphur; in the other, an energetic lime-washing of the houses and sheds will get rid of the annoyance.

It will be seen that by far the greater proportion of poultry diseases arise either from cold and wet, or neglect in preserving cleanliness—often both combined. It should be noted also, that the first general symptom of *nearly* all such diseases is diarrhœa, which we have observed usually manifests itself even in roup, before any discharge from the nostrils is perceptible. At *this stage* much evil may be warded off. Whenever a fowl hangs its wings, and looks drooping, let it be seen at once whether it appears purged, and if so, give immediately, in a table-spoonful of warm water, a tea-spoonful of strong brandy saturated with camphor. Repeat this next morning, and in most cases the disease, whatever it is, will be checked ; care being of course taken to give the invalid warmth and good shelter, with ale in its food. If the evacuation continues, administer the stronger prescription given for diarrhœa.

We could easily fill a long chapter with further prescriptions, but we believe that the above are all that can be *usefully* given. Special diseases, such as white comb in Cochins, and black-rot in Spanish, will be mentioned under the head of the breeds to which they more particularly belong.

SECTION II.

THE BREEDING AND EXHIBITION OF PRIZE POULTRY.

SECTION II.

On a subject involving so many conditions for success, and dependent so much upon circumstances, as the breeding of poultry for exhibition, it will be easily understood that the opinion of even the best authorities on some points is by no means uniform.

Many breeders, for instance, consider it almost a sin to try the effect of a cross; whilst others aver, with good reason, that crossing has done much towards the formation of some of our best breeds.

All, however, are agreed with respect to the essentials of practical rearing, and the following pages embody the experience and knowledge of the most eminent breeders in the kingdom. What *can* be taught by perusal we believe will be found here contained; and we trust this Section will be found of some real use in imparting information on matters concerning which nothing in any connected form has hitherto been written.

THE
BREEDING AND EXHIBITION OF PRIZE POULTRY.

CHAPTER VII.

YARDS AND ACCOMMODATION ADAPTED FOR BREEDING PRIZE POULTRY.

WHETHER the breeding of poultry with a view to exhibition can be made profitable, or otherwise, is a much vexed question amongst amateurs. For ourselves, we believe that the answer must depend partly upon the means of the fancier; still more upon the experience and knowledge he brings to bear upon the subject; and not a little upon the breed to which his fancy inclines him. We are acquainted with breeders who never could make the produce of their yards *quite* meet the current expenses; and we also know at least half-a-dozen, of high standing at all the principal shows, whose yards yield them a clear profit varying from £20 to £200 per annum. It is, therefore, most certainly *possible* to make even the "fancy" for poultry remunerative; and with the kind assistance of some of its most enthusiastic devotees we shall in this and the following chapters endeavour to give such information on the subject as can be thus communicated, and such directions as the long experience of many has proved likely to lead to success. But first of all it is necessary to consider the question of accommodation.

The plan of a poultry-yard given at page 11, with the

addition of a lawn or separate grass-run, on which young chickens may be cooped separately, is very well adapted for rearing most breeds upon a moderate scale. The two runs may be used to separate the sexes during autumn if preferred, or to keep the chickens apart from the old fowls; whilst the run for the sitting hens will, after this design has been fulfilled, be very convenient for the reception of one or two single cocks, or any other casual purpose. To ensure success, the most exquisite cleanliness must be observed, and at the beginning of every year the grass in the runs should be carefully renewed, if necessary, by liberal sowing, of course keeping the fowls off it till thoroughly rooted again. At this season the confinement thus involved will not be injurious, provided green food be supplied in the sheds, in lieu of the grass to which the birds have been accustomed. With such precautions, at least forty or fifty chickens may be reared annually, and from such a number there should be little difficulty, if the parents were selected with judgment, in matching two or three pens fit for exhibition.

But more extensive accommodation will be necessary if *very* high and extensive repute in any particular breed be desired, with the capability—which alone makes such reputation remunerative—of being able to supply an extensive demand for eggs and stock. In that case provision has to be made for keeping not only *separate strains*, in order that the proprietor may be able to cross and breed from the produce of his own yards, but there will be a much larger number of cockerels than can be needed, and as they are much too valuable for the table, they also have to be accommodated apart from the other fowls, until disposed of. We shall, by the kind permission of the eminent breeders whose establishments are represented, give two plans, each excellently adapted to secure these objects, though of very different arrangement; and which may easily e modified to meet any possible case.

The first (Fig. 12) represents the poultry-yard of Mr. H.

Fig. 12.

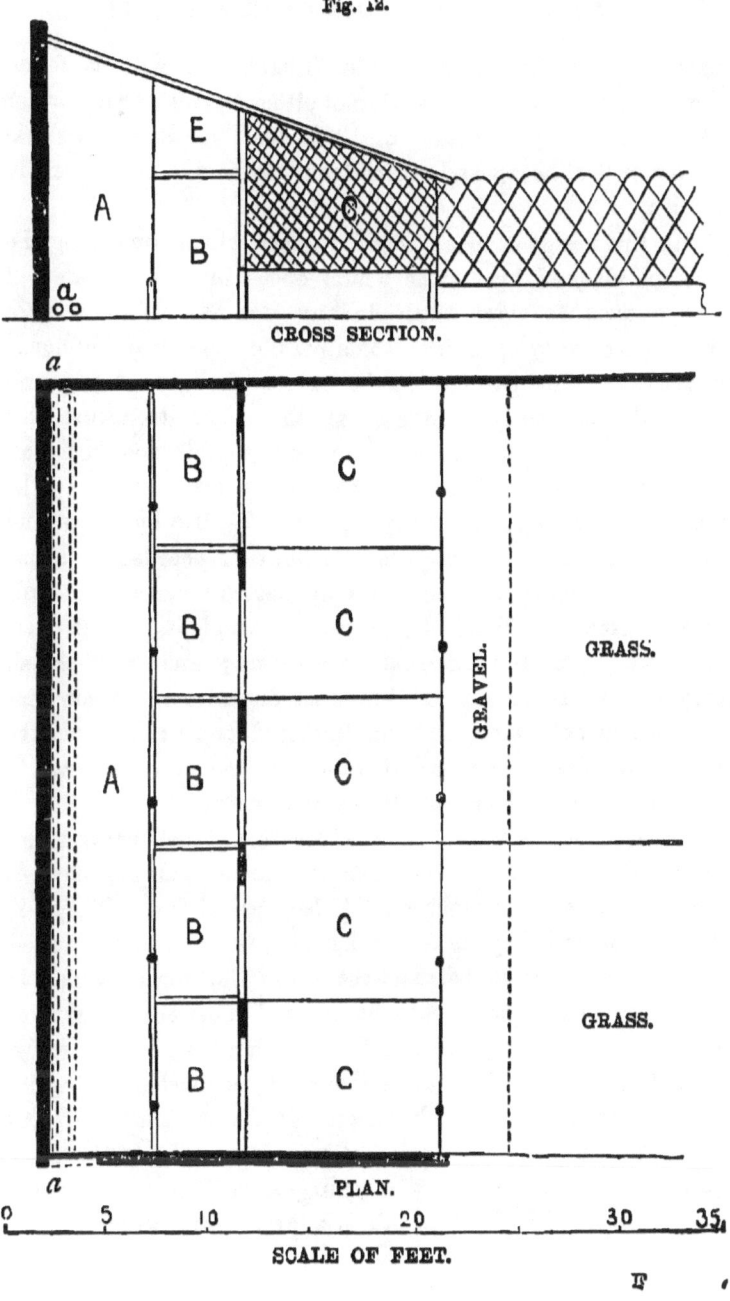

CROSS SECTION.

PLAN.

SCALE OF FEET.

Lane, the well-known fancier of Bristol, and will be found peculiarly adapted for the rearing of either Spanish or any other delicate breed; protection from inclement weather, as well as convenience of access and superintendence, having been specially studied.

In this design A is a covered passage which runs along the back of all, and by a door which opens into each allows of ready access to every house in any weather. One end of this passage may open into some part of the dwelling-house if desired. The passage should have a skylight at top, and must also be freely ventilated at the *roof;* to secure this object by having it open at either end would cause draught, and destroy the peculiar excellence of the arrangement. The houses, B, for roosting and laying in are $7\frac{1}{2}$ feet by 4 feet, and the side facing the passage is only built or boarded up about 2 feet, the remainder being simply netted; hence the birds have a free supply of the purest air at night, whilst quite protected from the external atmosphere; and can be all inspected at roost without the least disturbance—a convenience of no small value. The nests should be reached from the passage by a trap-door, and there is then no necessity ever to enter the roosting-house at all except to clean it.

A small trap-door as usual, which should be always closed at night, communicates between the house and the covered runs or yards, C, which are $7\frac{1}{2}$ feet by 9 feet. They are boarded or built up for 2 feet 6 inches, the remainder netted, except the partition between them and the houses, which is, of course, quite close. Both houses and runs must be covered with some deodoriser, and Mr. Lane prefers the powdery refuse from lime works, which costs about 1d. per bushel, and which he puts down about 2 inches deep. It always keeps perfectly dry, and is a great preventive of vermin; whilst if the droppings are taken up every morning, it will require renewal very rarely. In front of all is a grass run, which should extend as far as possible, and on which the fowls are let out in turn in fine weather.

An additional story, E, may or may not be constructed over the roosting-house, and in case of emergency, by sprinkling the eggs, may be made to accommodate sitting hens; but is not to be preferred for that purpose, for reasons given in Chapter IV. Every poultry-keeper, however, knows the great utility of such pens on various occasions which continually arise, and they will be found excellent accommodation for sick or injured fowls.

In Mr. Lane's establishment hot-water pipes ($a\,a$) are laid along the back of the passage floor, by which the temperature is at all seasons kept nearly uniform. This may or may not be adopted; and it will also be obvious that the whole arrangement is capable of enlargement to any desired extent.

Our second plan is of totally different design, and represents the yard of R. W. Boyle, Esq., of Bray, Co. Wicklow, Ireland.

In this design A A are roosts and enclosed runs adapted for breeding pens; the roosts in the larger pair measuring $11\frac{1}{2}$ feet by $6\frac{1}{2}$ feet, with a run extending 12 feet in front; in the smaller, the houses and runs are only 8 feet wide. B B are houses and runs adapted to receive either a single cock or pair of hens, and C C are still smaller for the same purpose, the roosts in the latter measuring 3 feet by 4 feet, and the open runs 4 feet by 6 feet 9 inches. Either of the latter, besides their specific purpose, are excellently adapted for the accommodation of a couple of sitting hens. D and E are large roosts or houses, which may be used to receive hens with their chickens, or for water-fowl. A grass plot, F, occupies the central portion of the yard, with a pond for the water-fowl. The parts lettered G are hard gravel. The entrance to the whole at H opens upon a large grass run, to which the fowls

68 BREEDING AND EXHIBITION OF PRIZE POULTRY.

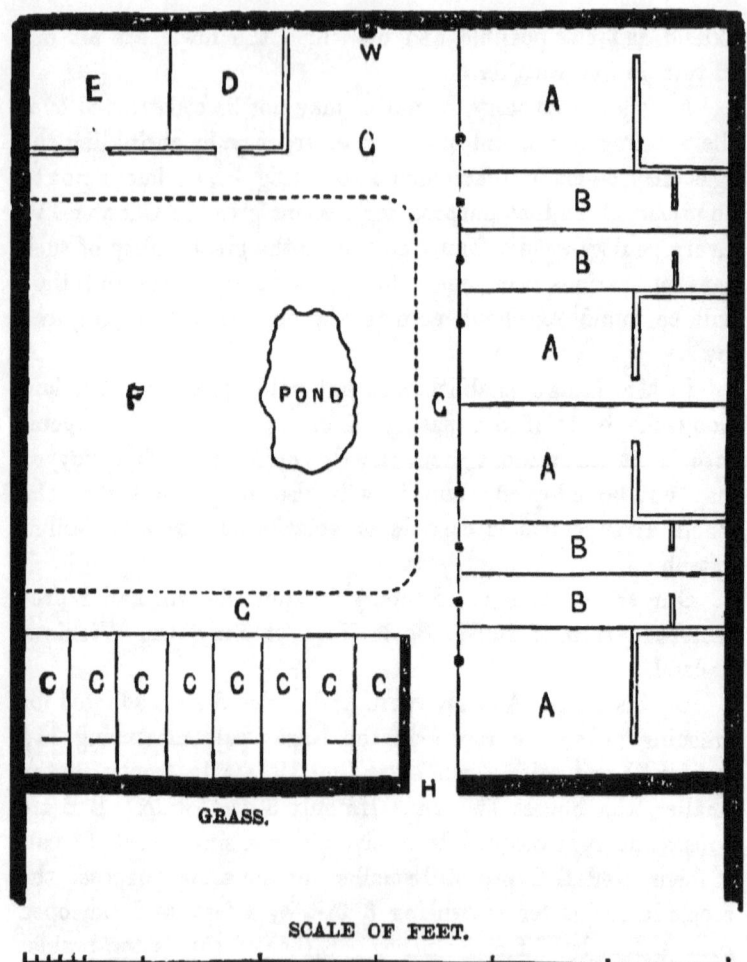

Fig. 13.

AA Roosts and Yards for Breeding Fowls.
BB Roosts and Yards for single Cocks or two Hens.
CC   Ditto,   ditto.
DE Houses without Runs.
F Grass Plot.   GG Gravel Walks.
H Entrance to large Grass Run.
W Watercock.

are admitted in turn. At W is a water-cock for the general supply of the yard.

All the roosts and runs in Mr. Boyle's yard are well covered with loose sand, which is raked clean every morning; and the large grass run outside is furnished with a long shed for shelter, and a small house with nests for such hens as prefer to lay there.

Prize poultry may be also reared most successfully, and with very little trouble or expense in accommodation, in a park or on a farm. All habitual frequenters of shows must have observed the remarkable *constitution* exhibited in Lady Holmesdale's poultry; and we paid, by invitation, a visit to Linton Park, specially to learn the management which produced such excellent results, and to enjoy a chat with Mr. J. Martin, the well-known superintendent of the Linton poultry-yard. We found the system most simple, and to all who have equal space at command, the least expensive that can possibly be. Stone houses with gravelled yards there certainly are, but these were unoccupied by a single one of the Dorkings for which the Viscountess has obtained so wide a reputation, and Mr. Martin keeps practically the whole of the stock at perfect liberty in the park. Portable wooden houses are employed, mounted on small wheels, and without a bottom, which are placed in sufficiently distant localities to avoid any danger of the birds mixing, and moved a little every two or three days. Open windows are also provided, so that the fowls always breathe the pure air of heaven, and certainly with much more freedom than most breeders would allow to such delicate varieties as Spanish and Dorking; yet Mr. Martin finds both breeds become *hardy* under such treatment, and that many of the Spanish birds prefer to roost on the trees, even through the winter. The hens are set in single detached coops, roofed on top, and closed at back and sides, which are placed in any secluded spots amongst the trees. Under this management the chickens are reared with the greatest ease, the gloss on the plumage is exquisite, and its closeness approaches that of the game fowl; whilst the birds, never too fat for the highest health, are surprisingly heavy in the scales.

A similar plan may be pursued on a farm; a number of wooden portable houses being provided, and placed in separate fields, in which families may be kept. Such a system will be an actual benefit to the soil, and the only drawback is the facility it affords to the felonious abstraction of valuable eggs and stock. Still, even with this objection, we must pronounce such a *natural* method of rearing far the best where it can be adopted, which is, however, in very few instances; for farmers are only seldom poultry-fanciers, and usually look upon even ordinary fowls as an unprofitable drain upon their purses, though it is certainly their own fault if it is so.

The intending prize-winner must, of course, adapt the plan of his yard to his own circumstances and situation. We have now given ample materials to furnish a design of any possible character. The one necessity in this class of poultry-keeping is *some* facility for what may be called separation or selection, combined, of course, with a healthy run for the chickens whilst young, and the essentials mentioned in the first chapter. If these can be secured, any design, with care and attention, and good breeding stock, will ensure a fair measure of success.

CHAPTER VIII.

ON THE SCIENTIFIC PRINCIPLES OF BREEDING, AND THE EFFECTS OF CROSSING.

To obtain any marked success in Poultry Exhibition it is very necessary that the scientific theory of breeding for any specific object should be thoroughly understood—at least, if anything like *general* eminence be expected; and still more so if the fancier desires by his own exertions to render any special service by the addition of new varieties, or the improvement of the old. Distinction in any one single breed is not so difficult to obtain; but he is a poor poultry-breeder who is content to let his favourite variety remain exactly as he found it, without at least some attempt to improve it either in beauty or in economic value; and any such attempt, to be successful, must be directed by an intelligent mind, which sees definitely before it the result to be attained.

In knowledge and enterprise of this description we cannot but confess that English fanciers are behind their Continental brethren; and the fact is the more to be regretted since the poultry "fancy" is far more universal in this country, and much more time and money spent in its pursuit. Were breeding more scientifically *studied*, no one can say what results British enthusiasm and perseverance might not eventually produce; whilst as it is, from ignorance of the subject, we believe one breed at least (white-faced Spanish) to have been nearly ruined. The elements of success are moreover so very few and simple, and a thorough knowledge of them so quickly acquired and so easily applied, that we shall devote a few pages to this part of the subject before entering upon the more practical portion of this section.

The greatest misapprehension appears to exist amongst all but the most educated poultry-fanciers respecting the origin of different breeds. People seem to imagine that they have come down to us, or at least a number of them, in unbroken descent from far-back ages; and this belief has given rise to innumerable discussions concerning the purity or otherwise of different varieties, which might have been spared had the disputants comprehended the real nature of the case. We cannot do better here than give some able remarks which appeared some time since in *The Field;* and which deserve to be well studied, for they contain the first principles of the whole science of breeding:—

"Such questions as the following are constantly asked,— 'Are the Brahmas a pure breed? are black Hamburghs a pure breed?' &c., &c. Those queries obviously owe their origin to a confusion of the distinction that exists between different animals, and between different varieties of the same animal. Let us illustrate our meaning by an example.

"A hare is a pure-bred animal, because it is totally distinct from all other animals, or, as naturalists say, it constitutes a distinct species. It does not breed with other animals, for the so-called leporines are only large rabbits; and if it did, the offspring would be a hybrid or mule, and almost certainly sterile, or incapable of breeding. In the same manner the common wild rabbit is a pure breed. This animal possesses the capability of being domesticated, and under the new circumstances in which it is placed, it varies in size, form, and colour from the original stock. By careful selection of these variations, and by breeding from those individuals which show most strongly the points or qualities desired, certain varieties, or as they are termed 'breeds' of rabbits, are produced and perpetuated. Thus we have the lop-eared breed, the Angora breed, the Chinchilla breed, &c. &c., characterised by alterations in the length of the ears, in the colour of the fur, in the size of the animals, and so on. It is obvious that, by care, more new varieties may be produced and perpetuated. Thus, by mating silver greys of different depths of colour, white animals with black extremities are often produced, and these have been perpetuated by mating them together. The breed so produced is known as the Himalayan variety, and, as it reproduces its like, is as pure and distinct a breed as any other that can be named.

"But, in the strictest scientific sense of the word, no particular variety of rabbit can be said to be a pure breed, as, like all the others, it is descended from the wild original. In the same manner we may deny applicability of the term pure breed to the varieties of any domesticated animal, even if, as in the case of the dog or sheep, we do not know the original from which they descended.

"All that can be asserted of the so-called purest-bred variety is that it has been reared for a number of years or generations without a cross with any other variety. But it should be remembered that every variety has been reared by careful artificial selection, either from the original stock or from other varieties.

"In the strict sense of the word, then, there is no such thing as an absolutely pure breed—the term is only comparatively true. We may term the Spanish fowl of pure breed, because it has existed a long period, and obviously could not be improved by crossing with any other known variety; in fact, its origin as a variety is not known. But many of our domesticated birds have a much more recent origin. Where were game bantams fifty years ago? The variety did not exist. They have been made by two modes: breeding game to reduce the size, and then crossing the small game fowl so obtained with bantams. Yet game bantams, as at present shown, have quite as good a title to a pure breed as any other variety. In fact, every variety may be called a pure breed that reproduces its own likeness true to form and colour.

"The statement that Brahmas, Black Hamburghs, Dorkings, &c., are pure breeds is meaningless, if it is intended to imply anything more than that they will reproduce their like, which a mongrel cross between two distinct varieties cannot be depended on doing. There is no doubt but that many of our varieties have been improved by crossing with others. The cross of the bull-dog thrown in and bred out again has given stamina to the greyhound; and although generally denied, there is no doubt but that the Cochin has in many cases been employed to give size to the Dorking. In the same manner new permanent varieties of pigeons are often produced, generally coming to us from Germany, in which country the fanciers are much more experimental than in England, where they adhere to the old breeds with a true John Bull tenacity."

Applying the above scientific and lucid remarks to the subject under discussion, it is now universally admitted by all who have studied the matter that every variety of the domestic fowl has originated in a wild bird still existing—the common Jungle Fowl of India, known to naturalists as the *Gallus Bankiva* of Temminck, or *Gallus ferrugineus* of Gmelin. To describe this bird minutely is unnecessary; it will be enough to say that, except in the tail of the cock being more depressed, it resembles very closely the variety known as Black-breasted Red Game. The assertion that all our modern breeds should be derived from this fowl may seem at first sight a large demand on our credulity; but such a fact is not more wonderful than that a cart-horse should have descended from the same original stock as the Arabian, or that an Italian greyhound and a Newfoundland should have common progenitors, about which no naturalist has the slightest doubt. The process is simple, and easily understood. Even in the wild state the original breed will show *some* amount of variation in colour, form, and size; whilst in domestication the tendency to change, as every one knows, is very much increased. By breeding from birds which show any marked feature, stock is obtained of which a portion will possess that feature in an *increased degree;* and by again selecting the best specimens, the special points sought may be developed to almost any degree required.

A good example of such a process of development may be seen in the "white face" so conspicuous in the Spanish breed. White *ears* will be observed occasionally in all fowls; even in such breeds as Cochins or Brahmas, where white ear-lobes are considered almost fatal blemishes, they continually occur, and by selecting only white-eared specimens to breed from, they might be speedily fixed in any variety as one of the characteristics. A large pendent white ear-lobe once firmly established, traces of the white *face* will now and then be found, and by a similar method is capable of development and fixture; whilst any colour of plumage or of leg may be obtained and established in the same way. The original amount of character required is *very* slight; a single hen-tailed cock will be enough to give that characteristic to a whole breed; and the two laced pullets mentioned under the head of Brahmas in the next Section would be quite enough, in skilful hands, to lay the foundation of a new and beautiful variety.

Any peculiarity of *constitution*, such as constant laying, or frequent incubation, may be developed and perpetuated in a similar manner, all that is necessary being care and time.

That such has been the method employed in the formation of the more distinct races of our poultry, is proved by the fact that a continuance of the same careful selection is needful to perpetuate them in perfection. If the very best examples of a breed are selected as the starting point, and the produce is bred from indiscriminately for many generations, the distinctive points, whatever they are, rapidly decline, and there is also a more or less gradual but sure return to the primitive wild type, in size and even colour of the plumage. The purest black or white originally, rapidly becomes first marked with, and ultimately changed into the original red or brown, whilst the other features simultaneously disappear.

If, however, the process of artificial selection be carried too far, and with reference *only to one* prominent point, any breed is almost sure to suffer in the other qualities which have been neglected, and this has been the case with the very breed already mentioned—the white-faced Spanish. We know from old fanciers that this breed was formerly considered hardy, and even in winter rarely failed to afford a constant supply of its unequalled large white eggs. But of late years attention has been so *exclusively* directed to the "white face," that whilst this feature has been developed and perfected to a degree never before known, the breed has become one of the most delicate of all, and the laying qualities of at least many strains have greatly fallen off.

It would be difficult to avoid such evil results if it were not for a valuable compensating principle, which admits of *crossing*. That principle is, that any desired point possessed in perfection by a foreign breed, may be introduced by crossing into a strain it is desired to improve, and every other characteristic of the cross be, by selection, afterwards bred *out* again. Or one or more of these additional characteristics may be also retained, and thus a *new variety* be established, as many have been within the last few years.

A thorough understanding of both the foregoing principles is so important, that we shall endeavour to illustrate each by examples.

Without foundation by long-continued *selection* no strain can be depended on. For instance—the Grey Dorking is a breed which assumes within certain limits almost any variety of colour, and occasionally, amongst others, that now known as "silver-grey." By breeding from these birds, and selecting from the progeny only the silver-greys, that colour has been established, like any other might be, as a permanent variety, which breeds true to feather with very little variation. Now a pen of birds *precisely* similar in colour and appearance may, as at first, be produced from ordinary coloured Dorkings, and shown as silver-greys; and the most severe test may fail to discover any apparent difference between them and the purest-bred pen in the same show. But breeding would show the distinction instantly: whilst one pen would breed true to itself, and produce silver-grey chickens, the *accidental* pen would chiefly produce ordinary Dorkings, with very few silver-greys amongst them; and though *in time*, by continuing to select these, a pure strain would ultimately be established, for immediate purposes the pen, as silvers, would be worthless. We know this to have been the case, to the great disappointment of purchasers. Conversely, even well-established silver-grey Dorkings, if bred from indiscriminately, will, by degrees, lose their distinctive colour, and go back to the ordinary grey stock from which they first sprang.

The coloured Dorking also exhibits very plainly the operation of *crossing*. It is evidently the produce of a cross between the original white Dorking and the large coloured Surrey fowl, as is proved by the fact that whilst the white Dorking—long established—invariably bred the fifth toe as its distinguishing characteristic, the coloured variety was for many years most uncertain in that respect. Still the fifth toe was introduced, along with the shape and aptitude to fatten; and by careful selection the colour and size of the Surrey fowl have been retained, whilst the tendency to only one toe behind, introduced by the cross, has been effectually eradicated, and the grey Dorking now breeds in this particular as true as the white.

The same fowl has been undeniably crossed with the Cochin in order to gain size, which has been retained to the great benefit of the breed, whilst all disposition to feather on the legs has been entirely bred out again. Game, again, has been repeatedly introduced into Dorking strains in order to gain constitution.

In the same way, when a race of Game fowls has been reduced in size, strength, and ferocity, by long interbreeding through fear of injuring the strain, a cross of the large, strong, and ferocious Malay at once restores the defective points, whilst all evidences of it are removed in three or four generations.

Perhaps, however, the most "artfully contrived" bird, and the best example of both principles combined, is to be found in the well-known laced Bantams of Sir John Sebright. This breed was founded by *crossing* the old Nankin Bantam with Polish fowls whose markings had a well-defined laced character. Lacing was thus imported into the Bantam breed, and by careful *selection* was developed and rendered perfect, whilst by the same process the Polish crest was effectually banished. This much being already accomplished, as we are informed by his son,* a hen-tailed Bantam cock accidentally met with struck Sir John's fancy, and added *that* peculiarity to the strain, which has now been for many years firmly established, and breeds as true as any, though so extremely artificial in its original "construction."

The last example we shall mention is the breed known as Black Hamburghs, which has been "made" within the last few years. That it has been obtained by *crossing* the Hamburg with the Spanish is proved sometimes *too* plainly by the evident traces of "white face" still lingering even in prize specimens; but the evidence of the cross will soon by *selection* be entirely bred out, and the breed has already made good its claim to a distinct class at most shows. The advantages gained by the cross are great. The size of the bird has been increased, and we have the enormous egg-producing powers of the Hamburgh race with a larger egg, thus doing away with the weak point of that beautiful breed.

But, it may be said, if these principles are correct, it would follow that the power of the breeder is almost unlimited. And practically it is so : there *are* within certain limits hardly any bounds to what may be effected by the scientific experimentalist. That so little *has* been done is mainly because the principles themselves have been so little understood, and most fanciers have been content to go on with the established varieties as they are, without any attempt to modify or improve them. There is another reason in the utter want of attention in this country to anything but colour of plumage and other "fancy" characteristics; and we cannot but think that our Poultry Shows have to some extent, by the character of the judging, hindered the improvement of many breeds. It will be readily

* It is only right to say that for these facts respecting the Sebright Bantams we are indebted to " The Poultry Book."

admitted in *theory* that a breed of fowls becomes more and more valuable as its capacity of producing eggs is increased, and the quantity and quality of its flesh are improved, with a small amount of bone and offal in proportion. But, if we except the Dorking, which certainly *is* judged to some extent as a table fowl, all this is *totally* lost sight of both by breeders and judges, and attention is fixed exclusively upon colour, comb, face, and other equally fancy "points."

We cannot but deeply regret this. We have shown how readily beauty and utility might be *both* secured ; and we do earnestly hope that even these pages may have some effect in stirring up our poultry-fanciers to the improvement in *real value*, without by any means neglecting the beauty, of their favourite breeds. The French have taught us a lesson of some value in this respect. Within a comparatively recent period they have produced, by crossing and selection, four new varieties, which, although inferior in some points to others of older standing, are all eminently valuable as table-fowls ; and which in *one* particular are superior to any English variety, not even excepting the Dorking—we mean the very small proportion of bone and offal. This is really useful and scientific breeding, brought to bear upon *one* definite object, and we do trust the result will prove suggestive with regard to others equally valuable.

We should be afraid to say how much might be done if English breeders would bring *their* perseverance and experience to bear in a similar direction. We have not, however, the slightest doubt that a breed of any desired colour might in a few years be produced, combining the Dorking quality of flesh with the prolificacy and hardihood of the Brahma, of which the cocks should weigh 20 lbs., and the hens 15 lbs. each. Many will question this : we simply say, that no one has yet attempted it, and that no one will doubt its possibility who knows the weights which *have* been occasionally attained in some of our largest breeds, and who has examined carefully into the effects already produced by judicious selection and crossing. But to obtain such a result, it must be systematically sought, and this will never be till the seeking is systematically encouraged by committees and judges.

In what way this could best be done, it is scarcely our province to decide; we shall be only too satisfied if our remarks be in any degree the means of directing attention to the importance of the subject. We believe, however, that a special prize of some value, announced annually, for award to the best pen, either of any known or new breed, for economic purposes, would shortly produce fowls, well established as a variety, that would astonish many old poultry-fanciers. Agricultural Societies in particular might be expected in *their* exhibitions to show some interest in the improvement of poultry regarded as *useful stock*, and to them especially we commend the matter.

CHAPTER IX.

ON THE PRACTICAL SELECTION AND CARE OF BREEDING STOCK, AND THE REARING OF CHICKENS FOR EXHIBITION.

WE have in the last chapter treated of the more theoretical principles which the breeder may employ in the accomplishment of any desired end; we have now to consider those practical points which the poultry-keeper must keep in mind if he desires to attain success in competition.

It is quite certain that there is nothing so unprofitable as to commence "poultry-fancying" with inferior fowls; and as there are always numbers of unscrupulous individuals who endeavour to impose upon the unwary, special caution is needed in the purchase of the original stock. If the reader be inexperienced, he should, if it be possible, secure the assistance of some friend upon whose judgment he can thoroughly rely; failing this, he should endeavour, not only by studying the descriptions, but by frequenting good shows, and seeing and comparing the live birds themselves, to become acquainted with at least the main points of the breed to which his preference inclines. To buy of unknown advertisers is always a great risk, and it will generally be found more economical in the long run to apply, in the first place, to known and eminent exhibitors, whose character stands too high to admit the suspicion of any wilful deception. Such breeders, it is true, will generally demand high prices for really good stock; but then the stock *will be* good, which is by far the most important point. Birds may also be purchased at shows; but in this case, if it is intended to breed from a single pen, it should be ascertained whether or not the cock is related to his hens, and if so, he should be exchanged for one of another family. In any case, the greatest care should be taken that the birds chosen are of *pure race;* it should be remembered that mere appearance is not always sufficient, as we have shown in the last chapter; and it is therefore most desirable to know the pedigree also.

At the very outset the question occurs, What is the best age to breed from? and we have no hesitation in replying that, according to the testimony of nearly all the best authorities, it is better the ages of the cock and hens should vary. It seems also generally admitted that the strongest and best chickens are produced from a cockerel nearly a year old mated with hens twelve months older; but, unfortunately, the chickens of such parents invariably have a large proportion of cocks, and most breeders therefore prefer a two-year-old cock with well-grown pullets not less than nine months in age. It must not, however, be supposed that either rule is imperative, or that good chickens are not to be expected from birds all hatched about the same time. In this case, however, it is advisable that all the fowls should be fully twelve months old; if younger, the chickens are usually backward in fledging. Fowls are good for breeding up to the age of four years, but are of little value afterwards.

To avoid any *near* relationship is most important; but many works have laid far too much stress upon the necessity of continually introducing what they call "fresh blood." It is certainly most destructive to breed from members of the same *family*, and to go on promiscuously interbreeding in one yard is still worse; but if there be a number of separate runs, in which separate races can be reared, operations may be carried on for many successive years without a cross from any other yard. It is the more necessary to explain this, because when any strain has been brought to high excellence, the introduction of a bird from another is a very serious thing, and we have personally known, in more than one instance, to ruin the produce of a whole year.

The plan to be adopted is to note down most carefully the parentage of every brood, and to keep the chickens from one family together until they are required. The breeding-yards for next year are then to be made up from the best specimens, taking care not only that the cocks and hens are not related *inter se*, but that two *runs* at least are thus made up without any fraternal relationship between them. Unrelated chickens will thus be secured for next year also; and so the system can be carried on. It is also a good plan, where it can be adopted, to put a promising young cockerel out to "walk" at a farm, or in some brother fancier's yard, and bring him back in a year or two, when the relationship between him and the pullets of the year will be too remote to be of very much consequence.

If a bird is occasionally introduced from another strain— and it certainly is advisable now and then, especially in the case of Dorkings—we can only say that the extremest care must be taken to ensure he is of good pedigree, as well as a perfect specimen in outward appearance of the breed to which he belongs.

Long experience has ascertained that the male bird has most influence upon the *colour* of the progeny, and also upon the comb, and what may be called the "fancy points," of any breed generally; whilst the form, size, and useful qualities are principally derived from the hen. Now it cannot be denied that it is desirable to secure absolutely *perfect* birds in all respects of both sexes if possible; but alas! every amateur knows too well the great scarcity of such, and the above fact therefore becomes of great importance in selecting a breeding-pen. For instance, a cock may have been hatched late in the year, and therefore be decidedly under the proper standard in point of size, and inferior for a show pen; but if his colour, plumage, comb, and other points—whatever they may be—are perfect, and he be active and lively, he may make a first-class bird for breeding when mated with good hens. A hen, again, if of large size and good shape, is not to be hastily condemned for a faulty feather or two, or even for a defective comb, if not too glaringly apparent—though the last fault is a serious one in either sex. But a very bad coloured or faulty-combed cock, however excellent in point of size, or a very small or ill-shaped hen, however exquisite in regard to colour, will invariably produce chickens of a very indifferent order.

It is also to be observed, with regard to the crosssing of a breed, that the cockerels in the progeny will more or less resemble the father, whilst the pullets follow the mother. A knowledge of this fact will save much time in "breeding back" to the original strain, and much disappointment in the effect of the cross. For instance, if it be desired to increase *size*, a cross with a *hen* of foreign breed should be employed, and the same if it be sought to introduce a more prominent breast, or any other peculiarity of shape; but if it is the plumage which is to be modified, it is the *male* bird who should be thrown in. In breeding the cross out again, or in retaining any new characteristic, so as to form a fresh variety, the same rule must be kept in mind.

We believe that much disappointment and uncertainty in the results of crossing has been owing to a neglect or ignorance of this simple principle, and breeding from either sex indifferently. If this be done, the result will often be worthless, and in every case the time consumed will be much greater than is necessary; but if scientifically conducted, we believe crossing would improve many of our older breeds in size, hardihood, and utility, without in any measure detracting from those qualities for which they are valued.

The care and preservation in good condition of valuable fowls is an important point. With regard to mere health, nothing can be added to what has already been treated of in the preceding section. But it frequently happens that, on account of the high price, only a single pen of three first-class birds can be afforded; and if such a family be penned up by itself, the frequent attentions of the cock will soon render the hens unfit for exhibition, whilst the birds may also mope, for want of more companionship. To avoid this, a couple more of ordinary hens should be added, taking care that the eggs be of a different colour, or otherwise easily distinguished from those of the breeding-pen itself. The plumage of the hens or pullets will then be preserved, without injuring the character of the progeny. We should, however, prefer mating the cock with four good hens of his own breed,—a plan more really economical, as the cost of the cock, in proportion to the number of eggs for sitting, is thereby reduced.

The number of hens, if good size and vigour are desired, should not exceed four. Many breeders allow six; but the finest fowls of the larger kinds are bred from the proportion we have stated.

It is desirable also, as much as possible, to save the hens from the wear and tear of chickens, which often injure the plumage greatly. It will not answer to prevent them sitting altogether; we have already remarked that such a procedure often causes them to suffer in moulting, which should not be risked. Neither do we altogether approve of the plan followed by many, of allowing them to hatch, and then giving the chickens to other hens. This may be done, if necessary, but a better system, where there is convenience for it, is to set a valuable hen upon duck eggs. The ducklings will not only resort to the hen to be brooded much less frequently than chickens, but will be far earlier independent of her care, and leave her in much better condition than if she had hatched her own eggs.

With regard to hatching, it is desirable with the hardier breeds to get the eggs under the hen as soon after January as a sitter can be obtained, in order that the brood may have all the year to grow in, and be ready for the earlier shows. At this season, however, the limitation as to number, mentioned in Chapter IV., must be strictly enforced, and no hen given more than seven or eight eggs, six chickens being as many as are desirable, in order that they may be well covered by the hen when partly grown, which is their most critical period as exhibition fowls. Spanish, Dorkings, or other delicate breeds, should not be hatched till April or May, unless unusually good shelter is at command.

As eggs are often purchased for hatching, it is necessary to allude to the frequent disappointments experienced in this respect, and which are far too frequently attributed, in no measured terms, to fraud on the part of the seller. Now we certainly cannot deny that such fraud is only too common. We know of one case where the fact was put beyond a doubt by examination, proving that the eggs purchased from a well-known exhibitor were actually *boiled;* but we honestly believe that the great majority of breeders would scorn such proceedings. It should be remembered, in the first place, that highly-bred birds are seldom so prolific as more ordinary stock, and are generally rather too fat for full health and vigour. Too many eggs—the full dozen—are likewise very often set, at seasons when the hen cannot give them heat enough; so that all get chilled in turn, and disappointment ensues. Bad packing also causes its share of failures; and, lastly, eggs are sometimes kept a week or fortnight after receipt before setting, which is always, but *especially* after a railway journey, most injurious. We can only recommend—1. That a hen be *ready* for the eggs before they are ordered. 2. That they be procured from a breeder of known honour and probity. 3. That especial directions be given that they are well packed. 4. That they be put under the hen *immediately* upon their arrival. And 5. That in cold weather the eggs be divided, so as not to exceed the number stated under each hen.

Eggs are best packed in small baskets, with the top tied down. If in boxes, the covers should be tied down or *screwed*, not nailed on any account, or every egg will be endangered. The best packing is to wrap every egg carefully in a separate wisp of soft hay; then to wrap each so enclosed in paper, to keep the hay from slipping off; and, finally, to imbed the eggs, thus guarded, in hay cut into 2-inch lengths; chaff or bran is too solid. Eggs so packed will go hundreds of miles without injury.

The chickens being hatched, let the utmost care be taken of them in every way. The object in this branch of poultry-breeding is not, as in the last section, to get a profitable amount of meat with the least possible expenditure in food; but, the birds being presumably good in quality, to get them by *any* means to the greatest possible size. For although size is never the first point considered, except perhaps in the case of Dorkings, it not unfrequently gives the casting vote between two contending pens, and is itself a most desirable point in nearly every fowl. Game and Bantams may be excepted.

The best stock food is undoubtedly oatmeal, and for valuable chickens it should be used liberally. With respect to this part of the treatment, however, we will give at length the remarks of one of the most successful breeders of Brahmas (the largest variety of fowl known), whose birds have in point of size been usually beyond all competition, and who has most kindly described for this work the system which has had such satisfactory results. The same feeding is applicable in every case where size is a point of merit.

"If the chickens are early hatched, I coop the hen in a warm sheltered place, free from all intrusion, and should the weather be very severe keep them within doors; the floor, however, must be gravel. Till about a fortnight old I feed them on sops made with boiled milk, and sweetened with coarse sugar, mixing it for the first two or three days equally with yolk of egg boiled hard and chopped fine. The egg is, however, too "binding" to be continued longer. The first thing in the morning they have warmed milk to drink; there is nothing equal to this for bringing them on in cold weather If the chicks are weakly, yolk of egg beaten up and given to *drink* is the most strengthening thing I know. In water they are of course unlimited, and they also have plenty of fresh grass cut small. I also throw them two or three times a day a handful of coarse raw oatmeal.

"I feed like this, on milk sops, raw oatmeal, &c., with milk every morning, for about a fortnight, after which they have boiled oatmeal porridge made so stiff that it will crumble when cool. They grow amazingly fast on this food, and are very fond of it. I also give them boiled rice occasionally, and frequently throw them groats, giving them also a little fresh cooked meat at dinner-time, cut up fine. Of course they are fed every night, after dark, usually about ten o'clock. There is at first a little difficulty in getting them out to feed at night; but they soon learn the time, and will run out eagerly for their 'stirabout,' which, if made thick enough, they prefer to any other food. The mode of preparation is to boil a saucepan full of water, and throw in as much oatmeal as will take it all up. Then continue stirring till it is a stiff crumbly mass, after which turn it out upon a large plate and keep stirring it about with the spoon till cool enough to be eaten.

"At ten weeks old, all the waste birds should be picked out to make more room for the others, and the cockerels separated from the pullets. The main food will still consist of the porridge, with small tail wheat, good heavy oats, and *plenty* of green food. Good potatoes boiled and mashed are also excellent food for a change.

"A little camphor put in their drinking water will help very much to keep them in health."

We have little to add to the above remarks. We do not ourselves approve of giving bread sops so long, and feel sure after trial, that chickens get on better by substituting oatmeal after the first day or two, or indeed from the day they break the shell. In cold weather also, a little sulphate of iron, or "Douglas mixture" should always be added to the water, and a little bread soaked in ale will be found beneficial. The warm milk is excellent, and is much better than the plan recommended by many of giving custard; the latter is too pampering, and after it chickens will sometimes refuse plain wholesome food. For weakly chickens, however, it is most strengthing to mix up a raw egg with their oatmeal. Above all, unless they have a good run on grass, the supply of green food must be unlimited.

For prize chickens, it is a good plan to mix with their meal a portion of the various condiments known as "cattle food" or "cattle spices." The appetite is thereby increased, and in confinement the birds grow faster and keep in better condition.

Feed often—every hour, if possible, from daybreak, and let the food be always fresh—nothing but grain or *dry* meal being ever allowed to remain. With such treatment and good shelter, if the stock be good and the number has been judiciously limited, the hen will not fail to bring a fair proportion through the most inclement season, and they will be sure to reach a good standard in point of size, having the best time of the year before them when they really *begin to grow*.

It is necessary to give one more caution. Do not let prize chickens roost *too soon*—never before they are at least three months old—and then see that the perches are large enough, and not round on the top, but like the flat side of an oval. If they leave the hen before the proper age for roosting, let them have every night a good bed of nice clean dry ashes. We never allow our own chickens, even while with the hen, to bed upon straw : ashes are much cleaner, and if supplied an inch deep are warmer also. To this plan we attribute a very small proportion of losses, even in very severe weather.

If a good field or other grass run be at command, the chickens will of course have it, and it will go a long way in supplying *all* other defective arrangements. But to our own knowledge some of the finest and largest fowls we have ever seen have been reared in a gravelled yard, not more than eighteen feet square. In such circumstances, besides the most scrupulous cleanliness and good feeding in other respects, there must be green food *ad libitum*—really fine chickens *cannot* be reared without it, their plumage in particular being of a very inferior appearance, and quite devoid of that beautiful "bloom" which is now indispensable to success in the show-pen.

But with proper care, and attention to the above plain directions, there should be no lack in due season of good fine birds. As they grow, and get through their first moult, they will be anxiously scanned, and let the best have especial care, taking out for the table all which are manifestly not up to the mark, that the rest may have more attention. We have already said that the sexes should be separated. This is highly essential in the larger varieties to good size, as too early a call on nature degenerates the breed; and had it been acted upon earlier by exhibitors of poultry, we believe the standard of weight in most fowls would have been now considerably higher than it is. There will thus be secured also greater vigour and fertility during the breeding season. The cockerel should not be put with the pullets intended for exhibition with him, till a fortnight before the show, but it is desirable that the pullets should have a little longer to get used to each other if they have been previously separated.

With the special treatment in view of exhibition, however, we will begin another Chapter.

CHAPTER X.

ON "CONDITION," AND THE PREPARATION OF FOWLS FOR EXHIBITION; AND VARIOUS OTHER MATTERS CONNECTED WITH SHOWS.

CHICKENS are rarely fit for exhibition until at least six months old, or even more. If the cockerels and pullets have been separated, as recommended in the last Chapter, there will rarely have been any eggs laid before this time; and stimulating food should now be partially discontinued to retard their production as long as possible, bearing in mind that the commencement of laying almost, if not quite, *stops the growth*, which it is desirable to prolong as far as possible for exhibition birds. In this respect the fancier and the ordinary poultry-keeper proceed upon contrary principles, the one endeavouring to get his pullets into laying order as soon as he can, the other using every expedient to procure a precisely opposite result.

If the chickens have been from the very shell properly and systematically fed, they will, by the time they are fit for showing, be in quite as good condition as they ought to be. By giving them two or three times a-day as much *soft food* as they will eat, they may easily be got up to any degree of obesity; and such a system of feeding is necessary to success at some shows, where the judges seem ignorant of the proper condition of a really healthy fowl; but we must most emphatically raise our voice against the practice. Let it be remembered that birds so fattened are, comparatively at least, *for ever ruined* for breeding purposes; that few chickens will ever be hatched from them, and those few delicate and sickly; and the reflection may perhaps cause the breeder to hesitate before he sacrifices, it may be the best stock in his yard, to any exhibition shrine. We cannot too severely condemn the conduct of those judges, who by their decisions help to maintain such prejudicial practices, and thereby render practically *barren* many of the finest birds ever bred. We have known a splendid pen of Dorkings, far superior in *real* size, as measured by the framework of the fowl, passed by contemptuously because inferior in mere dead weight to a pen which it would have been hopeless to breed from. There are, however, honourable exceptions: the most eminent judge in England always refuses to award a prize to a pen which he considers over-fattened; and thereby does all he can to check a system which prevents many celebrated breeders from sending at all to shows where such practices are known to prevail.

What we consider—and our opinion is corroborated by the best judges—to be really "good condition," is such an amount of flesh as can be carried consistently with perfect health and fecundity, combined with clean, well-ordered plumage. It is in the last particular that a good grass-run is so advantageous; fowls always look clean and nice when so kept, and rarely require much further preparation beyond washing the feet and legs.

With a good number of such birds to choose from, there

should be little difficulty in "matching a pen," even for Birmingham or Manchester. Matching is a matter of no small moment, as bad selection is fatal. Each bird is of course supposed to be of a fair good size, and tolerably perfect in form, colour, and feather. The two hens must then be carefully examined and compared with each other. Let it first be seen that the colour of their legs, eyes, and plumage generally corresponds, and that their combs and general proportions are alike also. Then let every part be examined in detail, seeing that the neck-hackle, back, and tail are the same in colour and marking; then the breast and wings. If all be satisfactory, and the birds be up to the mark, they should have a good chance of winning.

And let them not be judged *too* severely. Let the owner remember that few birds are absolutely perfect; and that whilst he, well knowing every fault, may see most plainly the blemishes in his own pen, impartial judges often have to weigh other blemishes against these, and he may thus win after all. Glaring faults cannot of course be passed over; but fair general excellence will often win the day against a pen far superior in some respects, if accompanied by some decided blemish.

The pens should be matched and the birds put together at least ten days before the show prepared for, in order that the fowls may get thoroughly used to each other. Neglect of this precaution may cause much fighting and destruction of plumage in the exhibition pen, or on the road thither, and not unfrequently loses a prize.

For the following observations on preparation for and sending to exhibition, we are indebted to Mr. F. Wragg, the well-known superintendent of the poultry-yard of R. W. Boyle, Esq. When it is remembered that this gentleman's fowls have always to undergo a sea voyage from Ireland, in addition to the ordinary railway journey, previous to exhibition, the beautiful "bloom" and condition in which they invariably appear, will cause his remarks to be appreciated by all amateurs.

"The system I pursue previous to sending to shows is as follows :—About a week beforehand I select the pen I intend to send, seeing, of course, that they match well, and carefully wash their heads and legs. I then have a nice dry room pretty thickly covered with clean straw, in which I put them, scattering a few handfuls of wheat amongst it. They scratch the straw about searching for the grains, and thus clean themselves beautifully without further trouble. The birds being kept up by themselves get so used to each other they never quarrel, either on the journey or in the pen. They have to drink clean water with a little sulphate of iron dissolved, which causes a bright red colour in the ears and comb, and makes them look well and sprightly.

"They are fed on oatmeal and Indian meal well boiled together, with a small quantity of salt just to season it; when properly done it is like a thick jelly. Twice, however, during the week, not more, they have rice, which is prepared by adding 1 lb. to a pint of water, and boiling till the water is absorbed, then adding as much milk as it will take up without getting thin, with a handful of coarse brown sugar; keep stirring the whole till done, and then put in a bowl to cool. Of this they are very fond, and it keeps them from purging. I also give them plenty of fresh green food.

"In their hamper I put, of course, plenty of clean soft straw. I also tie on one side of it, near the top, a fresh-pulled cabbage, and on the other side a good piece of the bottom side of a loaf, of which they will eat away all the soft part. Before starting I give each bird half a table-spoonful of port wine, which makes them sleep a good part of the journey. Of course, if I go with my birds, as I generally do, I see that they, as well as myself, have "refreshment" on the road.

"With regard to what you have remarked about showing

birds fat, I never do so. As you truly observe, many birds are ruined by it. Good, healthy condition, with a nice gloss on the feathers, is what I aim at in exhibiting, and the treatment I have described is what I have found best calculated to attain it."

Little can be added to these directions from so high an authority. For white fowls, however, or which have much white in their plumage, the cleansing process above described will often be found insufficient. In such cases the birds must be carefully *washed* with soap and water the night before sending off. Take a fine sponge, and, having well soaped it, smooth *down* every feather repeatedly, so as to clean without ruffling it; then repeat the process with water only till the soap is removed, and, lastly, with a soft towel. Let the birds be then left for the night in a *box* well littered with clean straw, open to, but not too near, the fire. Soda should never be used, as it stains the feathers yellow; and even the soap must be mild, without much free alkali. If they have had an extensive run on grass, however, the whitest fowls scarcely ever need washing, except as regards their feet and legs, giving also attention to the comb and wattles, if necessary. It is the poor dwellers in towns who have to take such precautions, and have so much to contend against. Yet, in spite of all this, we often see town breeders beating the very best country yards; and the fact proves that care and good system are of even more importance than any mere natural advantages.

Many exhibitors recommend the giving of linseed for a week before exhibition. Its use is to impart lustre to the plumage, which it does by increasing the secretion of oil. The fowls generally refuse the seed whole, and the best method of administration is to add a small portion of the meal daily to the ordinary soft food. A preferable plan, however, and one which agrees better with the health of the fowls, is to let the evening repast of grain for the last fortnight consist of buckwheat and hempseed in equal portions, which will be equally effective, and is greedily devoured by the birds, adding also to the beauty of the combs and wattles. We recently exhibited, at a first-class show, a pen of dark Brahma chickens, which took the first prize. The redness of the combs and the exquisite gloss on the plumage—every feather shining like velvet—were much admired; and we have repeatedly been asked the means by which such condition was attained. The only secret was the use of hempseed and buckwheat, with "Douglas mixture" (see page 30) in the drinking water, combined, of course, with good feeding generally.

Much difference of opinion exists as to the best form of hamper, but general experience approves most of a round shape, of a size to give just ample room to the fowls which have to be shown. Square corners are apt to catch the tails, and cause damage. For Spanish or other large-combed breeds it is best to have no cover, simply stitching a strong piece of canvas over the top; but for most fowls a wicker top is best, as affording more protection. It is of some consequence to committees that these covers should be *flat*, in order that the baskets may be compactly stowed away in the exhibition-hall.

In cold weather let the hamper be well lined with canvas, or straw stitched to the wicker-work. And if occupied by geese, let *special* care be taken that their bills cannot reach either the string fastenings or the direction-labels. They have a peculiar fancy for breakfasting upon those articles; and even fowls will occasionally contract the same vicious habit.

All has now been done that can be done, and the rest must be left to the decision of the judges. It is but rarely that fault can be found with their verdict: their duties are most arduous, and the manner in which, as a rule, they discharge them is deserving of the highest praise; but one or two are known to have certain invincible prejudices, which prevent them from judging some classes in accordance with the general rules as understood by the majority. This is to be regretted, as it hinders the good understanding which always ought to exist between judges and exhibitors. The object of both ought to be identical—the promotion of the highest standard obtainable in the different breeds, but it is necessary to this that the breeder should know definitely and authoritatively *what* he is to seek after. The "Standard of Excellence" did good service here, and was much wanted, but it is silent on many points, and, with reference to some others, is avowedly ignored by many judges. We think there is much need for a revised and larger work on the same basis;* and, in the meantime, it is our opinion that exhibitors have decidedly a *right* to know beforehand *who are to judge* their birds. To call upon them to send their best stock to a show where, it may be, the judges' *known* prejudices on certain points give them no chance of a prize, is evidently unfair.

But we are leaving the fowls, and must return to them, though we have little more to add. Whether they require any special treatment on their return will chiefly depend upon the system of feeding which has been pursued during the period of exhibition. If, as is too often the case, the pernicious plan of feeding on whole barley *ad libitum* has been retained, the birds will all be more or less feverish and disturbed, and will need a corrective. But such feeding cannot be too strongly condemned. It saves trouble certainly, but if a committee are not willing to take so much pains as will keep the birds in perfect health, they have no right to gather them together. We have the highest possible authority for saying that the best feeding is either barleymeal or oatmeal in the morning, mixed rather dry, and given before the public are admitted, with *wheat* in the

* Very complete scales of points, founded on actual analysis of modern judging, have, since these remarks were written, been published by the author in "The Illustrated Book of Poultry."

evening; and, in each case, only as much as the fowls will eat at once, without leaving any in the pens. Only these two meals should be given, as the birds have no exercise, and do not require more, besides which, the natural excitement of the show is best counteracted by a rather spare diet. Water should be given three times a day for a short time only—say five or ten minutes—not left for the birds to drink at will. Barley ought not to be used at all, as it is next to impossible it can be properly digested.

Fowls fed as here recommended will be returned in as good condition as they were sent, and require no attention at all beyond seeing that they do not get too much water and green food *at first*. But if they return from a "barley-fed" show, or the system on which they have been fed is unknown, or, in any case, if they appear either feverish or "overdone," give each a rather scanty meal of stale bread-crumb soaked in warm ale, let them have two or three sips *only* of rather tepid water, and then administer a tea-spoonful of castor-oil to each bird. This will probably be at night. Next day feed them on meal only in moderation, see that they cannot drink to excess, and give them half a cabbage-leaf each, or a large sod of grass, but no other green food; afterwards let them return to their usual diet. It is in all cases safest not to let them have any grain, and to put them on an allowance of water for the day after their return.

If our recommendations be attended to, there will be little injury from exhibition, and the same birds may be shown again and again without suffering. We know of fowls which have won as many as *fifty* prizes; and indeed *first-class* exhibition birds are almost always shown pretty frequently. They want care and attentive examination after each competition to see that they are not losing health; if it appears so, whatever other engagements may have been made, let them have *rest* till completely recovered; otherwise, property worth scores of pounds may be sacrificed for "just one more cup," to the owner's lasting regret.

We know not that we can usefully add any more upon this part of the subject. Something *must* be learnt by experience, for which no written directions can be substituted; nevertheless, we are not without hope that these few pages may prove of service in guiding the reader through the, perhaps, hitherto untried ordeal of the exhibition hall.

SECTION III.

DIFFERENT BREEDS OF FOWLS:
THEIR CHARACTERISTIC POINTS, WITH A COMPARISON OF THEIR MERITS AND PRINCIPAL DEFECTS.

WHITE COCHINS.

DIFFERENT BREEDS OF FOWLS.

CHAPTER XI.

COCHIN-CHINAS OR SHANGHAES.

THE Cochin breed, as now known, appears to have been imported into this country about the year 1847; those so-called exhibited by Her Majesty in 1843 having been not only destitute of feathers on the shanks, but entirely different in form and general character. No other breed of poultry has ever attracted equal attention, or maintained such high prices for such a length of time; and the celebrated "poultry mania," which was mainly caused by its introduction, will always be remembered as one of the most remarkable phenomena of modern times. To account in some measure for this, it should be remembered that no similar fowls had ever been known in Europe; and when therefore Cochins were first exhibited, it was natural that their gigantic size, gentle disposition, prolificacy, and the ease with which they could be kept in confinement, should rapidly make them favourites with the public. But the extent to which the passion for them would grow no one certainly could have foreseen. A hundred guineas has repeatedly been paid for a *single cock*, and was not at all an uncommon price for a pen of really fine birds. Men became mad for Cochins, and spent small fortunes in procuring them; and all England, from north to south, seemed given over to a universal "hen fever," as it was humorously termed. Their advocates would have it that the birds had *no* faults. They were to furnish eggs for breakfast, fowls for the table, and better morals than even Dr. Watts' hymus for the children, who were from them "to learn kind and gentle manners," and thenceforward to live in peace.

Such a state of things, of course, could not last, and the breed is now perhaps as unjustly depreciated by many as it was then exalted; for Cochins *have* great and real merits, and on many accounts deserve the attention of the poultry-keeper. The mania, absurd as it was, did however good service by awakening a general interest in the whole subject of poultry, which has never since died out.

As now brought to perfection, the breed presents the following characteristics :—

The cock ought not to weigh *less* than 10 or 11 lbs., and a very fine one will reach 13; the hens from 8 to 9 or 10 lbs. The larger the better, if form and general make be good.

The breast in both sexes should be as broad and full as possible; the general want of breast being the greatest defect in this breed. The neck can hardly be too short in either sex, so that it does not look clumsy; and the back must be short from head to tail, and very broad. The legs to be short and set widely apart, and the general make to be as full, wide, and deep as possible.

The shanks are profusely feathered down to the toes, and the thighs should be plentifully furnished with the fine downy feathers denominated "fluff." The quality of this "fluff," and of the feathering generally, is often a pretty good indication of the breed : if fine and downy, the birds are probably well bred; but if rank and coarse, they will not be worth attention as *fancy* birds. There is a tendency in the cocks to scanty furnishing on the thighs; but the breeder should choose a bird with as much "fluff" as he can get; not, however, allowing vulture hocks, which often accompany the heaviest feathered birds, but which are now disqualified at all first-class shows. The colour of the shanks is yellow, a tinge of red being rather a recommendation than otherwise; but green or white legs are to be avoided.

The head should be neat and rather small; the comb of moderate size, straight, erect, and evenly serrated: a notched or twisted comb is a great blemish. The ear-lobes must be pure red, no white being allowed. The eye ought in colour to approach that of the plumage, and should appear bright and sprightly.

The tail of the hen is very small, and nearly covered by the feathers of the saddle, which are very plentiful, and form a softly rising cushion on the posterior part of the back; the tail of the cock is larger than in the hen, but still small, and must not be very erect, or contain much quill; the wings in both sexes very small, neatly and closely folded in, and the general carriage noble and majestic.

The principal colours now bred are white, buff, and partridge. The white and buff are most popular.

The white must be perfectly pure in every feather; and green legs, which are apt to occur in this colour, will disqualify any pen, however meritorious otherwise.

In buff the colour may be any shade, but all the birds in a pen must correspond; black is admissible in the tail of both sexes, but the less there is the better. Black pencilling in the hackle is very objectionable, and a bird so marked will have no chance at a *good* show; but a little marking, if well defined so as to form a slight necklace, with no trace of indistinctness or clouding, is not to be regarded as a *fatal* fault. The colour of the cock should correspond with the hens on the breast and the lower parts of the body; but his hackles, wing coverts, back, and saddle hackles, are usually a rich gold colour. It should be observed that buff birds generally breed chickens lighter than themselves, and that most birds get rather lighter each moulting season; the breeding stock should therefore be chosen one or two shades darker than the colour desired.

In partridge hens the neck hackles are bright gold striped with black, the rest of the body light brown pencilled with a very dark shade of the same colour; the cock's hackles and saddle bright red striped with black, back dark red, wings the same, crossed with a sharply defined bar of metallic green black; breast and under part of the body black, not mottled.

Black used also to be shown, but has nearly disappeared, from the almost impossibility of keeping the colour free from stain. The other colours are grouse and cinnamon. The latter is well described by its name; grouse is merely very dark partridge. Cuckoo Cochins are never correct in form, and we believe are produced by crossing with the Gueldres.

The *merits* of Cochins have already been hinted at. The chickens, though they feather slowly, are *hardier* than any other breed except Brahmas, and will thrive where others would perish; they grow fast, and may be killed when twelve weeks old. The fowls will do well in very confined space, are very tame and easily domesticated, and seldom quarrel. They cannot fly, and a fence two feet high will effectually keep them within bounds. As sitters and mothers the hens are unsurpassed; though they are, unless cooped, apt to leave their chickens and lay again too soon for very early broods. Lastly, they are prolific layers, especially in winter, when eggs are most scarce.

Their *defects* are equally marked. The flesh is inferior to that of other breeds, though tolerably good when eaten young; there is, however, always a great absence of breast, which excludes the fowl from the market, and confines it to the family table. The leg, which contains most meat, is, however, providentially not so tough as in other breeds. The want of breast is best overcome by crossing with the Dorking, the result being a very heavy and well-proportioned table fowl, which lays well, and is easily reared. The hen, excellent layer though she is, has also an irresistible inclination to sit after every dozen or score of eggs; and this is apt to be very troublesome, except where a regular and constant succession of chickens is desired, when it becomes a convenience, as broods can be hatched with the greatest regularity. Finally, this breed is peculiarly subject to a prejudicial *fattening*, which, if not guarded against by the avoidance of too much or too fattening food, will check laying, and even cause death.

Cochins are subject to an affection called white comb, consisting of an eruption on the comb and wattles much resembling powdered chalk, and which, if not dealt with in time, extends all over the body, causing the feathers to fall off. The causes are want of cleanliness and of green food, chiefly the latter. This must, of course, be supplied, with an occasional dose of six grains of jalap to purge the bird; and the comb anointed with an ointment composed of four parts cocoa-nut oil, two of powdered turmeric, and one of sulphur.

On the whole, we consider this breed most useful to supply the *family* demand for either chickens or eggs, or to provide sitters for numerous broods; but it is little valued as a market fowl unless crossed with the Dorking or Crèvecœur; neither will it be found profitable where eggs are the sole consideration, and the hens cannot be allowed to indulge their sitting propensities.

CHAPTER XII.

BRAHMA POOTRAS.

IT is not our province to enter at length into the long disputed and still unsettled question as to whether Brahmas originated in a cross with the Cochin, or are entitled to rank as a distinct variety. There is much to be said on both sides. In favour of the Cochin cross may be named the gigantic size, the feathered legs and general appearance, the colour of the eggs, and formation of the skull; whilst those who believe it distinct have strong arguments in the altogether unique and peculiar comb, the colour, the prominent breastbone, the very different disposition and habits, and the opinion of, we believe, every eminent breeder. But one thing is certain: ever since this magnificent breed was introduced, it has steadily become more and more popular, and is now one of the most favourite varieties. To prosper thus, in the total absence of any poultry "mania," a breed must have real and substantial merits. Such Brahmas unquestionably have; and we shall endeavour, therefore, to give that full description of them which both their high rank as economic poultry, and their rapidly growing popularity, alike demand.

Their most marked peculiarity is in the comb, which is totally different from that of any other variety. It resembles *three* combs pressed into one. In a first-class cock, the effect is such as would be produced were a little comb, about a quarter of an inch in height, laid close to each side of his own proper comb, twice as high, the centre one being thus higher than the others. Each division of the comb ought to be *straight* and even, irregular or twisted combs being serious faults in a show-pen. In the hens the comb is very small, but the triple character should be equally evident, and the formation is quite plain even when the chicks first break the shell.

When first introduced, single-combed Brahmas were often shown, but are now scarcely ever seen, and rarely take prizes if there are any decently good pea-combed birds at the same show.

There are two varieties of Brahmas exhibited, known as "Light," and "Dark" or "Pencilled" Brahmas; and on no account should they ever be crossed, the result being, according to Mr. Teebay, who was formerly the most successful and extensive breeder of Brahmas in England, always unsatisfactory.

FEATHERS.

No. 1 is a *Striped* Feather.
,, 2 a *Laced* Feather.
,, 3, 4 are *Spangled* Feathers, No. 3 being from a Golden "Yorkshire Pheasant," and No. 4 from a Lancashire "Mooney" Hen.
,, 5, 6, 7, 8 are *Pencilled* Feathers, No. 5 being plucked from a Hamburgh, and 6, 7, 8 from a Dark or Pencilled Brahma.

The cross may be known, if the birds profess to be "dark," by the lighter, gayer appearance of the cock's breast, perhaps accompanied with large white splashes, and sandy coloured or brownish *patches** about the pullets. Should the fowls be offered as "light" Brahmas, the pullets will have buff, yellowish, or sandy backs and wings, and the cocks most likely yellowish hackles.

The following description of light Brahmas has been carefully drawn up under the supervision of John Pares, Esq., of Postford, near Guildford, well known as the most eminent exhibitor of this variety for many years past :—

"Light Brahmas are chiefly white in the colour of the plumage, but if the feathers be parted, the bottom colour will often be found of a bluish grey, showing an important distinction between them and white Cochins, in which the feathers are *always* white down to the skin. The neck hackles should be distinctly striped with black down the centre of each feather. (See "Feathers," No. 1). That of the cock is, however, often lighter than in the case of the hen. The back should be quite white in both sexes.

"The wings should appear white when folded, but the flight feathers are black.

"The tail should be black in both sexes. In the cock it is well developed, and the coverts show splendid green reflections in the light. It should stand tolerably upright, and open well out laterally, like a fan.

"The legs ought to be yellow, and well covered with white feathers, which may or may not be very slightly mottled with black: vulture hocks are a great defect.

"The ear-lobes must be pure red, and every bird should, of course, have a perfect pea-comb, though good birds with a single comb have occasionally been shown with success."

* This must not be confounded with the brownish tinge which nearly all " dark " Brahma hens acquire with age.

The "dark" or "pencilled" Brahmas are similar to the above in comb, form, symmetry, &c., but as different in colour as can well be. By the kindness of R. W. Boyle, Esq., of Bray, Ireland, who has for some years been known as the most eminent breeder of dark Brahmas in the United Kingdom, we are enabled to give the best description of this magnificent variety which has ever been published, most carefully drawn up by him for publication in these pages.

"The head of a perfect Brahma cock should be surmounted by a good 'pea-comb,' which resembles three small combs running parallel the length of the head, the centre one slightly the highest, but all evenly serrated and straight, and the whole low and set firm on the head. Beak strong, well curved, and the colour of horn. Wattles full: ear-lobes perfectly red, well rounded, and falling *below* the wattles.

"His neck should be rather short,* but well curved, with very full hackle, which is silvery white striped with black, and ought to flow well over the back and sides of the breast. At the head, the feathers should be white. Back very short, wide, and flat, rather rising into a nice, soft, small tail, carried rather upright. The back almost white. The saddle-feathers white, striped with black, as in the neck, and the longer they are the better. The soft rise from the saddle to the tail, and the side feathers of the tail, to be pure lustrous green black, except a few next the saddle, which may be slightly ticked with white: the tail feathers themselves pure black.

"The breast should either be black, or black with each feather slightly and evenly tipped with white, but on no account *splashes* of white: it should be well carried forward, full, and broad. Wings small, and well tucked up under the saddle-feathers and thigh fluff. A good sharply-defined black bar across the wing is very important.

"The fluff on the thighs and hinder parts ought to be black

* The "Standard" says *long*. A great error.—*Note by Author.*

DARK BRAHMAS.

or *very* dark grey. The lower part of the thighs should have plenty of nice soft feathers, almost black, rounding off about the joint and hiding it, but on no account running into 'vulture hocks,' which I consider a great eyesore.

"The cock should carry himself upright and sprightly, and great width and depth are important points: a good bird should *show* great size, and 'look big.'

"The hen's head should be small, with a perfect pea-comb, as in the cock, but smaller; and the beak also resembling his in the decided curve and colour. Wattles quite small and neatly rounded, the red ears hanging below them. Neck short, and gradually enlarging from head to shoulders. Feathers about the head greyish, verging to white, and the hackle more striped with black than in the cock.

"General make of the back, tail, thighs, wings, and breast, the same as in the cock, but of course in proportion.

"The colour of the hen, except the neck and tail, is the same all over, each feather, even up to the throat on breast, having a dingy white ground, very much and closely pencilled with dark steel grey. The pencilling on the throat and breast is very important, and is one of the first points looked at in a prize hen.

"The hen's legs are short and thick, not quite so yellow as the cock's, and profusely feathered on the outside with feathers the same colour as the body. Her carriage is scarcely so upright as that of the male bird.

"With regard to the economic merits of Brahmas, the pullets lay when six months old, and usually lay from thirty to forty eggs before they seek to hatch; but I have repeatedly known pullets begin to lay in autumn, and *never stop*—let it be hail, rain, snow, or storm—for a single day till next spring. I have kept several breeds, such as Dorkings, Spanish, and Hamburghs; but never now give to my tenants any but Brahmas, as they say they can rear them so much more easily, and greatly prefer them.

"As to their size, I cannot agree with those who think 'breeding for colour' detracts necessarily from this point. I have had a cock weighing fifteen pounds, and hens twelve pounds, but these are very unusual weights. I have, however, two cockerels of this year (1866), only six and a half months old, one of which weighs ten and three quarter pounds, and the other eleven and a quarter pounds. The latter I weighed off a grass run. He is the largest for his age I ever bred, and I am confident he will next year weigh fifteen to sixteen pounds, or even more. I consider twelve to thirteen pounds for a cock and nine to ten pounds for a hen very good weights. Cockerels for exhibition, when six months old, ought to weigh from eight to eight and-a-half pounds, and pullets from six to seven pounds.

"In breeding, it is necessary to be *very sure* the stock for generations back has not been crossed. I would then select the most perfect cock I could procure at any price, not less than twelve pounds weight, and mate him with either three pullets, or three hens a year old: if hens, to weigh at least nine pounds; if pullets, eight pounds. Each bird to be entirely free from vulture-hocks or brown-red feathers. From such parentage there will be little disappointment."

Mr. F. Wragg, the manager of Mr. Boyle's yard, adds the following valuable practical remarks on the breeding of Dark Brahmas:—

"I would on no account breed from birds with faulty combs, or the slightest twist in the tail, as such defects are most surely transmitted to a large proportion of their progeny. I would also reject a cock with 'splashes' of white on his breast, or a hen with *very* dingy brown in the feathers.

"I select, if possible, a cock with *perfectly black* breast, thighs, and fluff, and other qualities well defined, two years old, and twelve pounds in weight. I would put him with three pullets, their first season, square-built, short-legged birds, with broad-striped hackles, small and perfectly straight combs, and perfect in feathering. By this I mean that *each feather* should be most distinctly pencilled; and I am most particular that on the breast especially every single feather right up to the throat should show four or five distinct half circles of black on the same ground colour as the rest of the body. Let the pullets be nine pounds weight. Breed from such birds, and nearly half the chickens will be fit for exhibition.

"I wish to repeat, that for *breeding* I select a cock with all the underparts perfectly black. For exhibition, either the same colour or a little white mottling will do. A mottled cock looks best. I especially dislike to see the 'fluff' on the cock's thighs with white in it. Many of the chickens from such a parent would be very bad in colour, showing light 'streaky' feathers on the breast."

Joseph Hinton, Esq., of Hinton, near Bath, one of the earliest breeders of Dark Brahmas, adds a few remarks which also deserve attention.

"I have always striven," he says, "to keep Brahmas from being considered birds of colour only. The chief point in judging should be form, then size, then comb and colour. Body to be as broad and deep as possible : legs stout and wide apart, and cannot be *too* short, or too well feathered. The leg feathering ought to be abundant from the very hock. To see a nearly bare shank, even with a well-feathered foot, is very unsightly. I prefer a slight *tendency* to vulture-hock—that is, an abundance of soft curling feathers, projecting over the hock and hiding the joint: a naked hock to me is an abomination. Knock-knees also, which frequently occur in cockerels, are very objectionable.

"As to colour, I prefer myself the *clear* grey, but it is unfortunately liable to lightness on the breast. For this reason many breeders prefer a reddish-brown breast, but I myself should object to the reddish tinge.

"It is also objectionable when the flight or primary quills in the cock's wing are not well tucked under the outside part of the wing, though I think they have laid too much stress on this point in the 'Standard of Excellence.'* The fault is rarely seen in the master cock of the yard, and I believe it therefore to occur from the efforts of the junior birds to save themselves from punishment by the 'king of the walk.' In such struggles the wing is rapidly extended, and then often not fairly returned. In time this becomes a habit, and greatly mars the beauty of the bird."

The latter fault alluded to is unfortunately too frequent. It can, however, be cured by carefully returning every feather to its proper place, and then tying twine round the end of the wing, to prevent the bird from opening it till the feathers are re-set into their position. About a month will ensure this; and in the meantime the bird must of course sleep on straw, as it cannot fly.

Mr. Hinton's remarks on colour lead us to almost the only disputed point in this breed. Mr. Lacy, and other eminent breeders, avowedly *prefer* a decided brown ground colour for the hens, for reasons which we will give in his own words :—

"I have been a breeder of Dark Pencilled Brahmas," he says, "for fifteen years, ten of which I kept them in America, where they are the favourite fowls amongst farmers and planters. I began by purchasing the best I could find, which were beautifully symmetrical in shape, and very large, the cock weighing thirteen pounds, and the hens nine pounds each. The colour of the hens was as follows : neck-hackle white, streaked with black; saddle and wings a beautifully pencilled

---

* We rather agree with the "Standard" in this matter. The first prize cock at Birmingham last year (1866) had the defect alluded to, and the award of the judges was condemned by every breeder at the show on that very ground, though the bird was very fair otherwise.

brown, the ground colour being the *dark*, with lighter markings of a quarter-moon shape on each feather; breast a light salmon-coloured ground, with *dark* pencillings of the same quarter-moon shape, forming the most beautiful contrast of the two colours imaginable. The fluff had also the brown tinge.

"This colour I have striven to produce and sustain in my strain of birds, breeding as they do much more true to colour than the grey variety. This last, I believe, has been introduced by some cross, as I have obtained grey pullets from other yards whose produce has been mixed, whilst they themselves have moulted to the brown shade and sometimes even to the red. Besides this, the great difficulty of producing *light-breasted pullets* cannot be got rid of in grey birds; because, having, as I believe, been crossed with a lighter colour, they will 'throw out' a majority of inferior birds. I do not, however, like a *reddish-brown* colour; nor will any breeder be troubled with it, provided he uses proper discretion in the selection of his breeding stock."

Others maintain that the brown colour referred to is a blemish, and we must ourselves side with this view. But, whether we are right or wrong in this, it is certain that the variation in opinion is most unfortunate; for the difference of colour does not at all appear in the cocks, and hence there is always great danger, in purchasing a male bird, of injuring the pencilling which may be preferred. Each school, however, has a right to its own fancy, and we can only advise the utmost care in every introduction of fresh blood that may be made, that the *tinge of the strain purchased* corresponds with that already in the yard. It is to neglect of this precaution so many bad coloured, mottled, and "streaky" birds owe their origin.*

* The above remarks refer to Brahmas as they are now exhibited and judged. But we must remark that the birds formerly shown so successfully for several years together by Mr. R. Teebay, at Birmingham, and

"Vulture hocks" have also occasioned considerable discussion. The "Standard of Excellence" states that they are to be considered objectionable, but not a disqualification. Many breeders *defend* them, as being always more or less associated with heavy shank-feathering; but all first-class judges at present seem agreed to absolutely disqualify any pen in which vulture-hocked birds appear, though soft *curling* feathers tucked in nicely round, and hiding the joint, are decidedly to be preferred.

The *precision* of the pencilling is very important, on the breast especially, but has hitherto been overlooked in every published description of Brahmas. Every feather should be *distinctly* pencilled across several times with black, as are the pencilled Hamburghs, but more minutely, on a dull white ground. On the breast the marking should be equally distinct and abundant, but it there follows the outline of the feather, and becomes a series of four or five "lacings," one within the other. By the kindness of an eminent exhibitor and breeder of this variety we are enabled to give engravings of actual feathers taken from very perfect prize birds, which will illustrate this. (*See* plate of "Feathers," frontispiece.) No. 6 is a feather from the centre of a pullet's breast; No. 7 is from the flat of the wing; No. 8 from the coverts of the tail. Birds thus pencilled are of exquisite beauty, but second-rate specimens

many other shows, were much *darker* than now, the dark pencilling being so dense and black as to have quite a metallic *green* shade, which we have not now seen in hens for some time. The pullets are probably bred lighter through selecting cocks as free as possible from any red or bronze in the wing coverts, some amount of which appears essential to breeding dark birds. We simply *note* this change to a lighter shade as one too important to pass over; whether it be of itself any deterioration is, of course, a fair subject for difference of opinion. But many experienced breeders will also note changes in shape, and other characteristics—the result of various crosses, and which certainly are not improvements.

often show a cloudy, indistinct mass of minute and confused markings, which are far inferior in appearance.

At a show held at Oswestry last year (1866) a pen of Brahmas was shown in the "Light" Class, of which the two pullets were beautifully *laced* on the breast, with all the precision of a Silver Sebright Bantam. The effect was very pretty indeed, and we hope the variety may be perpetuated.

Little more need be added. With regard to the merits of Brahmas, they must certainly rank very high. In size the dark variety surpasses every other breed yet known, the heaviest cock ever recorded, so far as we are aware, having attained the enormous weight of *eighteen pounds*, and thirteen and fourteen being not uncommon at good shows; though only *good* strains reach this weight, and miserable specimens are often seen which are inferior in size to Cochins. They also lay nearly every day, even in the depth of winter, and if *pure bred*, scarcely ever sit till they have laid at least thirty or forty eggs. When they sit more frequently, the hen will usually be very brown, and is, we believe, crossed with the Shanghae. As winter layers, no breed equals them. We are writing at the end of November, and have a hen which has laid forty-five eggs in forty-eight days, whilst others are little inferior. Brahmas are likewise very hardy, and grow uncommonly fast, being therefore very early ready for table; in which particular they are profitable fowls, having plenty of breast-meat. They bear confinement as well as Cochins, being, however, far more sprightly; and scarcely ever, like them, get out of condition from over-feeding.

The flesh, however, though better than that of Cochins, is much inferior, after six months, to that of the Dorking; and this is their only real fault; but a cross with a Crevècœur or Dorking cock produces the most splendid table fowls possible, carrying almost incredible quantities of meat of excellent quality. Such a cross is well worth the attention of the farmer.

On the whole, there is no more profitable fowl "all round" than the Brahma; and a few hens at least should form part of the stock of every moderate yard.

CHAPTER XIII.

MALAYS.

THE Malay was the first introduced of the gigantic Asiatic breeds, and in stature exceeds that of any yet known. The cock weighs or should weigh from nine to eleven pounds, and when fully grown should stand *at least* two feet six inches high. But the general size of this breed has of late greatly deteriorated.

In form and make Malays are as different from Cochins as can well be. They are exceedingly long in the neck and legs, and the carriage is so upright that the back forms a steep incline. The wings are carried high, and project very much at the shoulders. Towards the tail, on the contrary, the body becomes narrow—the conformation being thus exactly opposite to that of the Shanghae. The tail is small, and that of the cock droops.

The plumage is very close, firm, and glossy, more so than that of any other breed, and giving to the bird a peculiar lustre when viewed in the light. The colours vary very much. We consider pure white the most beautiful of all; but the most usual is that well known under the title of brown-breasted red game. The legs are yellow, but quite naked.

The head and beak are long, the latter being rather hooked. Comb low and flat, covered with small prominences like warts.

Wattles and deaf-ears very small. Eye usually yellow.* The whole face and great part of the throat are red and naked, and the whole expression "snaky" and cruel. This is not belied by the real character of the breed, which is most ferocious, even more so than Game fowls, though inferior to the latter in real courage.

Malays are subject to an evil habit of eating each other's feathers, a propensity which often occurs in close confinement, and can only be cured by turning them on to a grass run of tolerable extent, and giving plenty of lettuce with an occasional purgative.

The chickens are delicate, but the adult birds are hardy enough. They appear especially adapted to courts and alleys, and may not unfrequently be seen in such localities in London.

The principal merit of Malays is as table fowls. Skinny as they appear, the breast, wings, and merrythought together carry more meat than perhaps any other breed; and, when under a year old, of very good quality and flavour. They also make good crosses with several breeds. Mated with the Dorking they produce splendid fowls for the table, which also lay well; and with the Spanish, though both parents are long-legged, the result is most usually a short-legged bird of *peculiar* beauty in the plumage, good for the table, and, if a hen, a remarkably good sitter and mother. They have also been extensively crossed with the English Game fowl, in order to increase the strength, size, ferocity, and hardness of feather.

* The "Standard" says the eye should be fiery *red*, but this is most decidedly wrong. We should, of course, hesitate to state positively that a "red eye" has *never* been seen; but we do say we never saw one, and doubt if any one else ever did, at all events lately. An eminent breeder of this variety informed us that the native fanciers in India preferred a pearly or *white* eye; but that there also he knew on good authority the red eye was unknown, except in very rare instances. How the "Standard" came to give "red" eyes as a point, is a mystery.

The great drawback of Malays is their abominably quarrelsome disposition, which becomes worse the more they are confined. The hens are also inferior as layers to most other breeds; and on these accounts the pure strain is not adapted to general use, though useful in giving weight and good " wings" to other varieties of fowl.

CHAPTER XIV.

GAME.

No variety of fowl has been so enthusiastically cultivated by amateurs as the Game, and in none perhaps is there so much room for legitimate difference of opinion. The varieties are legion, and to describe every one would be hopeless, except in a work specially devoted to the purpose; we shall therefore only give descriptions of the leading breeds, as written for this work by Trevor Dickens, Esq.,* of London, one of the most eminent authorities in England on all points connected with the Game fowl.

"The Game cock, as the undisputed king of all poultry, requires more careful judging in regard to shape, than any other bird. The Brown-reds have long been most perfect in outline; but the following description will apply to a perfect bird of any breed.

"The beak should be strong, curved, long, and sharp; the comb single, small, and thin, low in front, erect, and evenly serrated; it is usually red, but sometimes darkish red. Head long and sharp, with the face and throat lean and thin. Earlobes small and red, never whitish. Neck long, strong, and

* Well known for his annotations on the breeds of Game in the *Poultry Chronicle*, under the signature of " Newmarket."

well arched; the hackle short, hard, close, firm, and broad in the feather. Back short, and very hard both in flesh and feather; broad at shoulders, narrow at tail, and rounded at the sides. Breast broad and very hard, but not by any means too lean or too full—the last would be useless weight; a good hard breast is most essential, as it is the most vulnerable part of the bird. The rump should be narrow, neat, and short, the saddle feathers close, hard, and short. Wings very strong, and of a just medium length, well rounded to the body, and carried neither high nor low, but so as to protect the thighs. Very long-winged birds are usually too long in the body, and short-winged birds too broad in the stern. Tail neither long nor short, but medium length, and carried erect to show good spirit, but not 'squirrel-fashion' over the back; it should be well 'fanned,' or spreading, and the sickle feathers of a good round full curve, and standing clearly *above* the points of the quill tail-feathers.*
Very long-tailed birds are soft and long-bodied, and short-tailed birds are too short-winged, and often have broad rumps. Thighs short † and very muscular, hard, and firm; placed well wide apart, and well up to the shoulders, in order to give a fine fore-hand and make the bird stand firm on his legs; which latter should be sufficiently long, but not too much so, and placed wide apart as the thighs. Spurs low down, long, sharp, and rather thin; a little curved upwards, and not turning in too much. Feet flat, broad, spreading, and thin; the claws and nails straight, long, and strong; the back claw especially long and flat to the ground, to give a firm footing. The whole plumage should be *very* close, short, and hard, with glossy reflections, and the

---

* Many breeders, especially in Yorkshire and London, prefer close or folded tails. But, as a rule, the well raised and spread tail shows more spirit, if not clumsy, which is of course bad.

† The "Standard of Excellence" says "*rather* short." This is decidedly not emphatic enough to denote the proper proportions in a good cock.—*Note by Author.*

quills or stems strong and elastic. Body in hand short and very hard, and the general carriage upright, quick, fierce, and sharp. The back is best rather curved, provided it be flat *crosswise*, and not hump-backed or lop-sided. Weight for exhibition, 4½ to 5½ lbs.; for the pit, not over 4½ lbs.

"The hen should correspond in form, but of course in proportion, *hardness* of flesh and feather, with shortness of body, being main points. Good hens generally become spurred, and such breed the hardest and best cocks. The proper weight of a hen is from 3 to 3½ lbs.

"A short or clumsy head, short or soft neck, long body, narrow shoulders, long thighs, legs set close together, loose or soft plumage, and especially what is known as a 'duck foot,' are serious defects. It should be remembered that a Game fowl is *always* judged mainly in reference to its fighting qualities, and anything which interferes with them is a fault in the bird.

"With respect to the varieties of Game, the sorts which take nearly all the prizes and cups are the Brown-red, Black-breasted Red, Silver Duck-wing Greys, and Piles, all which are cup-birds.

"The Brown-red is essentially *dark* in blood, the eyes being a very dark brown, with the comb and face inclining to a dark gipsy purple, and the beak dark also. Breast of the cock a red-brown, shoulders sometimes passing into a rich orange-red colour. Wing-butts of a dusky or dark smoky brown, and general colour a dark red. Legs dark iron-brown or blackish bronze, with dark talons. Hackle with dark stripes, and thighs like the breast. The tail a dark greenish black, and the wing is often crossed with a glossy green bar. The general colour of the hen is very dark brown, grained or pencilled with lighter brown; her neck-hackle a dark golden copper-red, thickly striped with dark stripes; and her comb and face darker than in the cock bird. Good hens are usually spurred, and their tail feathers show a slight curve.

DUCK-WING GAME.

The Brown-red breeds are most esteemed in the Midland Counties, and at the principal shows take most cups. They are also the favourite breed with sportsmen, and are best in shape of all; but like all the dark-combed varieties, are not such good layers as those with bright red combs.

"Black-breasted Reds are essentially *red*-blooded birds, the plumage being generally a bright red, rather deeper on the body than in the hackle. *Red eyes* are absolutely essential to good birds, all others being inferior and infallibly denoting a cross. The cock's wings are bright red in the upper part, and rich red chestnut in the lower, with a steel blue bar across; breast bluish black, with glossy reflections; thighs the same; tail greenish black, the feathers without much down at the roots. The comb and wattles of all Black-reds must be *bright* red, and the legs are usually willow colour in cup birds, though any leg will do if the birds are bright in colour, and have red eyes. The general colour of the hen is a rich red partridge-brown, with a red fawn-coloured breast, and reddish golden hackle with dark stripes; the cock's hackle also is striped *underneath*, but clear above. Spurred hens are the best, but are not so frequent as in the preceding variety.

"Silver Duck-wing Greys are purer in blood than the Yellow or Birchen Duck-wings, and are white-skinned when of pure breed. General colour of the cock, a silver grey; hackle striped with black underneath, but clear above; back a clear silver grey; breast either bluish black or clear mealy silver colour; wing crossed with a steel-blue bar, and the lower part of a creamy white; tail greenish glossy black. Hen a silvery bluish grey, thickly frosted with silver; breast a pale fawn-colour; neck-hackle silvery white, striped with black. The comb and face in both sexes are bright red. The legs may be either white, blue, or willow; but of course the whole pen must match, and white leg to silver feathering is certainly the most correct match. Willow is, however, most common in the legs, but least pure in blood ; the white or blue-legged birds being the true-bred Silver Duck-wings. Eyes should be red in Willow and Blue-legged strains, and yellow in Yellow and White-legged strains in all the Duck-wing Game fowls.

"The *Yellow Duck-wings* are similar to the above except in the straw-colour or birchen tinge, and the copper-coloured saddle. They have yellow skins, and willow or yellow legs. In this variety the cock's breast is always black, the hen's a pale fawn colour, whilst the silver hen often has a clear mealy or silver breast instead of fawn.

"Red eyes and willow legs are the only correct colours for prize Duck-wings. Bright red eyes and white legs for prize Piles.

"The colour called Piles consists, in the cock, of a bright red piled on a white ground, the hackle being red and white striped; the back is chiefly red, and the breast mostly white, but often with red markings; the tail should be white, but a few red feathers are not amiss; *black* in the tail, as seen in the Worcestershire Piles, is, however, very objectionable. The hens are red-streaked or veined on a white ground, the breast redder than the cock, and the tail white, with a few red feathers occasionally. The reddest Piles are the best birds, and prize pens should be selected with bright red eyes and white legs.

"*Whites* should have bright red eyes, and white legs are essential.

"*Black* Game fowls should have black eyes and bluish black legs—have won a few cups.

"*Dark Greys* ought always to have black eyes and legs. The hens are very dark.

"The original wild varieties of Game fowls are three :—(1.) The Black-breasted Red, with fawn-breasted partridge hens ; (2.) Brown-breasted Reds, with dark legs, and dark brown (not black) hens; and (3.) Red-breasted Ginger Reds with yellow legs, and the hens a light partridge colour. These three colours were probably reclaimed at a very early period, and are still found in India as wild birds. From them all the other colours were originally bred; the varieties hatching dark chickens from the brown or dark reds, and all others from the other two sorts. These varieties can be merely named, and are most conveniently classed thus, according to the colour of their chickens when hatched :—

Light Chickens.	Striped Chickens.	Dark Chickens.
1. Whites.	5. Black-breasted Reds.	10. Brown Reds.
2. Piles.	6. Red-breasted Ginger Reds.	11. Dark Greys.
3. Blue Duns.	7. Duck-wings.	12. Dark Birchens.
4. Red Duns.	8. Yellow Birchens.	13. Black.
	9. Mealy Greys.	

"There are also four other varieties not generally known, called Red Furnaces, Cuckoos, Spangles, and Polecats, making at least seventeen well-defined sorts of Game fowls; but besides these, there are at least twenty-seven named sub-varieties, or forty-four in all. To describe these in detail would be useless, and I shall only, therefore, add the following general remarks :—

"The best criterion of blood in all Game fowls is the *colour of the eyes*, a point which has been, strange to say, totally overlooked in every work on poultry hitherto published. *Black* eyes show dark blood, and the hens of such strains lay white eggs. *Red* eyes denote red blood, and lay pinkish eggs. *Yellow* or *daw* eyes lay yellowish eggs. These last are inferior in spirit to the others. Brown and bay eyes result from crossing different breeds.

"The only sorts of much use for fighting are those with black or red eyes, and the three varieties now usually employed are the Brown-breasted Reds, Dark Greys (which are strongest and hardiest of all), and Black-breasted Reds, with *white* legs and dark red eyes. The sorts which fight the *quickest* are, however, the Red Cheshire Piles, with bright red eyes and white legs, the Red-breasted Ginger Reds, with bright red eyes and yellow legs, and Whites, with white legs and bright red eyes; but they have not quite so much strength and power of endurance. The Black-breasted Reds with *willow* legs are generally too slow and soft for the pit, as are the Blacks also.

"The best layers are the Black-breasted Reds with willow legs, the hens being partridge colour; and Red Cheshire Piles with white legs. The worst layers are the greys, Dark Greys and Dark Birchens being worst of all. With the exception of these, Game fowls lay *remarkably* well, and in favourable circumstances will, I believe, surpass *any* breed. My willow-legged Black-breasted Red hens have averaged from 211 to 284 eggs per annum. To reach this, however, they will require a good run, but if well attended to, are always good layers. It is worth remarking that yellow and blue-legged birds generally lay best in all poultry.

"Game cock chickens should be shown *undubbed*; but at their first Christmas they become 'stags,' and should then have their comb and wattles taken neatly and closely off with a very sharp pair of scissors.

"Different varieties ought not to be crossed, but kept distinct. In breeding either for stock or exhibition, nothing is so necessary as to have a good proportion of cocks. There should be one to every six hens at least; and as in a large yard it is impossible, from their pugnacity, to keep more than one full-grown brood cock, there should be a good supply of fine young birds or 'stags' kept under him, and breeding with the hens, when all the eggs will be fecundated, and the chicks vigorous and healthy. This is the only way of breeding good stock from a large yard; and it is of course preferable, when practicable, to keep each cock to his own limited family of hens. Pullets ought never to be bred from at all, and should be kept away from the cocks, using their eggs for household purposes. Good old birds will always breed strong chickens, and in this breed it scarcely matters how old they are so long as they remain strong and healthy. The breeding pens should be selected with great care, not from the largest, but from the *best-shaped* and *strongest* birds. The more cock chicks in a brood the better, as it is always an evidence of strength and vigour in the strain; and the pullets, though fewer, are finer and handsomer birds invariably.

"Game eggs should not be hatched before the 21st of March, nor after the end of May. This breed is of warmer blood and stronger constitution than any other, and the chicks consequently hatch earlier, often breaking the shell at the end of the nineteenth day. As soon as they begin to fight, the cocks should be separated, and, if possible, put out to 'walk' at a farm; the pullets will rarely injure themselves, and their quarrels are only amusing."

To the foregoing remarks of Mr. Dickens—the best description of the breed ever yet published—we shall only add a few sentences on the general qualities of Game fowls. Their merits are many and various. In elegance of shape, in hardihood, in bold and fearless spirit, what can equal them? But besides these recommendations, they rank, as already stated, in the very first class as layers, provided only they have a good run; whilst for delicacy of flavour their flesh is confessedly beyond any comparison. They should never be fatted, being too impatient to bear the process; but if eaten just as taken off their runs are equal to the pheasant. They also eat little, and are therefore profitable fowls, whilst as mothers the hen is not to be equalled. She should not be given too many eggs, on account of her small size; but she will hatch her full complement, and when hatched will take good care of them, defending them against *any* foe to the last gasp. If there be cats in the neighbourhood commend us to a good Game hen.

There are, however, a few drawbacks. The size of both birds and eggs is small, which of itself makes them of little value as a market fowl, and in confinement the Game hen will by no means lay so well as Brahmas, Spanish, or Hamburghs. Their pugnacious disposition also disqualifies them for small runs, though not to the extent generally supposed.

On the whole, we should pronounce this breed the very one for a country gentleman, who can give his fowls ample range; and it will in such circumstances afford a constant and abundant supply of the most delicious eggs and meat to be obtained. Their good laying qualities may also recommend them to the farmer in some localities. But they cannot be considered a profitable breed for domestic purposes in general, or to those whose object in poultry-keeping is to supply the market with table birds.

CHAPTER XV.

DORKINGS.

THIS is a pre-eminently English breed of fowls, and is, as it always will be, a general favourite, especially with lady fanciers. The general predilection of the fair sex for Dorkings may be easily accounted for, not only by the great beauty of all the varieties, but even more perhaps by their unrivalled qualities as table-birds—a point in which ladies may be easily supposed to feel a peculiar interest.

The varieties of Dorkings usually recognised are the Grey or Coloured, Silver Grey, and White. We believe the White to be the original breed, from which the coloured varieties were produced by crossing with the old Sussex or some other large coloured fowl. That such was the case is almost *proved* by the fact that only a few years ago nothing was more uncertain than the appearance of the fifth toe in coloured chickens, even of the best strains. Such uncertainty in any important point is always an indication of mixed blood; and that it was so in this case is shown by the result of long and careful breeding, which

GREY DORKINGS.

has now rendered the fifth toe permanent, and finally established the variety.

In no breed is size, form, and weight so much regarded in judging the merits of a pen. The body should be deep and full, the breast being protuberant and plump, especially in the cock, whose breast, as viewed sideways, ought to form a right angle with the lower part of his body. Both back and breast must be broad, the latter showing no approach to hollowness, and the entire general make full and plump, but neat and compact. Hence a good bird should weigh more than it appears to do. It is difficult to give a standard, but we consider that a cock which weighed *less* than 10 lbs., or a hen under $8\frac{1}{2}$ lbs., would stand a poor chance at a first-class show; and cocks have been shown weighing over 14 lbs. This refers to the coloured variety. White Dorkings have degenerated, and are somewhat less.

The legs must be white, with perhaps a slight rosy tinge; and it is imperative that each foot exhibits behind the well-known double toe, perfectly developed, but not running into monstrosities of any kind, as it is rather prone to do. An excessively large toe, or a triple toe, or the fifth toe being some distance above the ordinary one, or the cock's spurs turning outward instead of inward, would be glaring faults in a show pen.

The comb may, in coloured birds, be either single or double, but all in one pen must match. The single comb of a cock should be large and perfectly erect. White Dorkings should have double or rose combs, broad in front at the beak, and ending in a raised point behind, with no hollow in the centre.

In the Grey variety the colour is not material, so long as the two hens in the pen match. The cock's breast may be either black or mottled with white; the hackle, back, and saddle are usually white, more or less striped with black; and the wing we like best to see nearly white, with a well-defined black bar across.

In the Silver Grey Dorking, however, colour is imperative. This variety, there is not the slightest doubt, was at first a chance off-shoot from the preceding, but has been perpetuated by careful breeding. Coloured birds will always occasionally throw silver-grey chickens, and such are sometimes exhibited as "bred" Silver Greys; but it is needless to add that disappointment is sure to ensue, unless the strain has been kept pure for many generations. The Silver Grey colour is as follows:— Cock's breast a pure and perfect black; tail and larger coverts also black, with metallic reflections; head, hackle, back, and saddle feathers, pure silvery white; and the wing bow also white, showing up well a sharply-marked and brilliant bar of black across the middle. A single white feather in the tail would be fatal. Hen's breast salmon-red, shading into grey at the thighs; head and neck silvery white striped with black, back "silver grey," the white of the quill showing as a white streak down the centre of each feather; wings also grey, with no shade of red; tail dark grey, passing into black in the inside. The general appearance of both birds should be extremely clean and aristocratic.

The white birds should be what their name implies—a clear, pure, and perfect white. There is generally in the cock more or less tendency to straw or cream colour on the back and wings, and we would by no means disqualify a really first-class bird in all other points on account of it; but it is decidedly a fault.

White Dorkings are usually much smaller than the coloured, which we believe to have hindered the popularity of this truly exquisite variety. It has often occurred to us that this defect might be easily remedied by crossing with the large Grey Dorking, and then breeding back; and on a recent visit to Linton Park we saw the experiment fairly commenced, with every prospect of success. A good white cock had been mated with some light-coloured hens, and out of the progeny there appeared six or seven pure white chickens, of very great merit. Two cockerels attracted our special attention; they were not six months old when we saw them, but they were fully up to the Grey Dorking standard of size, and we have not the slightest doubt, when full grown, would weigh at least 12 lbs. each, whilst in colour they were quite equal to their parent. We commend this method of increasing the size to all White Dorking fanciers. We have also known a cross tried with the White Cochin, but never saw a bird so produced that was fit to look at.

We cannot let the subject of size pass without alluding to the great obligations Dorking breeders are under to Mr. John Douglas. By careful selection of stock, and close attention, with probably the help of a cross, he succeeded in raising the standard of this breed at least 2 lbs. higher than had ever been known before; and the fowls he bred have never yet been surpassed.

It should be remembered that Dorkings degenerate more than any variety from interbreeding; and, if fresh blood be not introduced, rapidly decrease in size. They also suffer much from frequent exhibition, not bearing confinement well. We are, however, inclined to think that in some degree this arises from the vicious practice of over-feeding the birds, to increase their weight, before showing; and we cannot help expressing our *decided* opinion that judges should always disqualify such an over-fed pen, however meritorious otherwise. No less is due to the public, who not unfrequently purchase prize pens to breed from; for a pen in such a condition of unhealthy fatness is not only useless at the time, but can rarely be got into really healthy condition again. We have seen a really magnificent pen of Dorkings, whose lives we would not have given three weeks' purchase for; and which, even if they did survive, were irretrievably ruined for breeding, and ought therefore to have been condemned by the judges.

Dorkings are peculiarly subject to "bumble foot"—a chronic gathering, or abscess, probably first produced by the heavy birds descending on the ground from too high perches, but now it appears more or less hereditary in the breed; at least we have seen it repeatedly in fowls never allowed to roost high enough to cause it in this way, and which had the unrestricted run of a spacious park. We believe there is no remedy but to let the abscess grow to maturity, and then remove it surgically. The operation will be successful about once out of three times.

The great merit of Dorkings has already been hinted at, and consists in their unrivalled excellence as table-fowls. In this respect we never expect to see them surpassed. The meat is not only abundant and of good quality, surpassing any other English breed except game, but is produced in greatest quantity in the choicest parts—breast, merrythought, and wings. Add to this, that no breed is so easily got into good condition for the table, and enough has been said to justify the popularity of this beautiful English fowl. It should also be noted that the hen is a most exemplary sitter and mother; and, remaining longer with the chickens than most other varieties, is peculiarly suitable for hatching early broods.

The Dorking is not, however, a good layer, except when very young; and in winter is even decidedly bad in this respect. The chickens are also of very delicate constitution when bred in confinement, and a few weeks of cold wet weather will sometimes carry off nearly a whole brood; they ought not, therefore, to be hatched before May. But it is only right to say that when allowed unlimited range the breed appears hardy, and as easy to rear as any other, if not hatched too soon. At Linton Park, the chickens are all left with the hens at night, under coops entirely open in the front; and grow up in perfect health, whilst the old birds frequently roost in the trees. It is in confinement or on wet soils that they suffer, and the only way of keeping them successfully in such circumstances is to pay the strictest

WHITE-FACED BLACK SPANISH.

attention to cleanliness and drainage, and to give them some *fresh turf* every day, in addition to other vegetable food. With these precautions, prize Dorkings have been reared in gravelled yards not containing more than 300 square feet.

In fine, the breed is most valuable for the market, or as a general fowl, on a wide and well-drained range. But we cannot recommend it to supply the table with eggs, or as a profitable fowl to be kept in a limited space.

Our illustration is drawn from a magnificent pair of Grey Dorkings kindly lent for the purpose by Lady Holmesdale.

CHAPTER XVI.
SPANISH.

UNLIKE almost all other varieties, there really appears some reason for believing that this breed of fowls did originate, or at all events come to us, from Spain. It has, however, been long known and valued by amateurs in this country, and perhaps no other is so generally popular. This is no doubt partly owing to their truly aristocratic and haughty appearance, but no less also to their unrivalled large white eggs, which exceed in weight those of any other breed, except the lately introduced La Flèche, and are always sought after for the breakfast-table.

Of all the varieties of this breed now known, the white-faced Black Spanish is by far the most important, and the only one for which a special class is reserved at most poultry exhibitions; all others having to be shown in the class "for any other variety." Of this truly beautiful breed the following description has been given us, and subsequently most carefully revised by Mr. H. Lane of Bristol, well known for his magnificent strain, and who has probably taken more first prizes with his birds than any other breeder within a similar period:—

"The general carriage of Spanish fowls is of great importance. The cock especially should carry himself very stately and upright, the breast well projecting, and the tail standing well up, but not carried forward as in some birds. The sickle-feathers should be perfect and fully developed, and the whole plumage a dense jet black, with glossy reflections in the light. The hen should be equally dense in colour, but is much less glossy. Any white or speckled feathers, which now and then occur, are fatal faults.

"The legs should be blue or dark lead-colour; any approach to white is decidedly bad.* The legs in both sexes are long, but the fowl should be nevertheless plump and heavy. I consider a good cock for exhibition ought not to weigh under seven pounds: the hen a pound less; and I have had several excellent cocks which weighed eight pounds each. All Spanish fowls in really good condition are heavier than they appear to be.

"The comb must be very large in both sexes, and of a bright vermilion colour. That of the hen should fall completely over on one side, but the cock's comb must be *perfectly* upright, the slightest approach to falling over being fatal to him at a good show. The indentations also must be regular and even, and the whole comb, though very large, quite free from any appearance of coarseness. Any sign of a twist in front is a great fault.

"The most important point, however, is the white face. This should extend as high as possible over the eye, and be as wide and deep as possible. At the top, it should be nearly arched in shape, approaching the bottom of the comb as nearly

* It is singular that the old fanciers imperatively required these identical *bluish white* legs in prize birds; and legs of too dark a tint were often put in *poultices* to make them light enough!

as possible, and reaching sideways to the ear-lobes and wattles, meeting also under the throat. In texture the face ought to be as fine and smooth as possible. The ears are large and pendulous, and should be as white as the face. Any fowl with red specks in the face has not the slightest chance.

"With regard to Spanish fowls as layers, the pullets will generally lay when six months old, and I seldom get less than five or six eggs a week from each. My house is warmed,* which has, of course, some influence on a breed so delicate; but with this artificial aid, I find my pullets lay throughout the winter, as above.

"The great thing with the chickens is to keep them out of the damp. They scarcely ever get roup; but if not kept dry die away rapidly, no one knows how. They ought not, therefore, as a rule, to be hatched very early in the year, and one cock ought not to be allowed more than three hens, as the eggs are less fertile than those of most other breeds."

The following additional remarks on this fowl are compiled from information furnished us by various amateurs.

Spanish are judged most of all by the quantity and quality of the "face." If this be rough and "warty" so as to hinder the sight of the bird, or have any decided red mark, especially above the eye, or be much disfigured in the same region by feathers, the bird has little chance. Such feathers are often pulled out, but if thus "trimmed" a pen ought always to be disqualified, though it is almost universally done by exhibitors.

Like all other black fowls, coloured or even white feathers will occasionally happen. Such birds are hopeless to exhibit, and decidedly bad to breed from.

That the comb of the cock should be absolutely erect is most important, and many breeders, to secure this, place light wire frames, or "cages," over them, as soon as sufficiently developed to hold the wire in place: the combs are thus grown

* For plan and description of Mr. Lane's establishment, see Chap. VII.

straight, like cucumbers! But there will rarely be need for this, if the breeding-stock be of good constitution. The hens selected for breeding should therefore be carefully chosen with good *thick* combs, which spring up with some arch before they fall over the side of the head. Hens with combs that fall dead over will rarely breed strong-combed cockerels. The comb of both sexes should, however, get thin at the edge, or it will appear heavy and clumsy.

Mr. Lane has alluded to the delicacy of the chickens. During feathering, which is in this breed a very slow process, they require special care and most generous diet, or few will be reared. When full grown, however, they are a tolerably hardy fowl altogether, but always suffer much in moulting, and during very cold or damp weather.

In no breed is purity of race of so much importance as in this; and in introducing a fresh cock it is especially needful to see that both his appearance and his pedigree are quite satisfactory. One of the most eminent breeders in England informed us a few months since that all his chickens of the season had been ruined by the introduction of a fresh cock, whose face when purchased appeared perfectly white, but who had imported more or less red into every chicken hatched from him. There can be no doubt, however, that too close interbreeding has greatly injured the Spanish fowl, and that both size, constitution, and prolificacy have been sacrificed to the white face alone. Such a result is to be regretted; and as it is now becoming generally acknowledged and deplored, we may hope that it is not yet too late to get back some of the size and hardihood of the Spanish fowl as formerly known.*

* It was a subject of general remark that at the last Birmingham show (December, 1866) the Spanish fowls were larger and finer on the whole than had been seen for a considerable time; and at the Bristol show a month after there was confessedly the most splendid collection of this breed that had ever been seen, as was emphatically remarked by the judges. We hope the improvement may not be merely temporary.

The other principal varieties of Spanish are—the Minorca, or Red-faced Black; the White; the Blue, or Andalusian; and the Ancona, Grey, or Mottled breed.

MINORCA.—This breed resembles in comb, ears, shape, and colour of plumage, the white-faced breed, but considerably surpasses it in size; and, on an average, we consider the comb more largely developed; the legs are also shorter. A good cock ought to weigh from eight to nine pounds. It is the best layer of all the Spanish breeds, and the chickens are tolerably hardy. It is a great favourite in the West of England, and deserves to be more widely cultivated, as it far surpasses the preceding in everything except the white face. Prizes are now and then offered to Minorcas, and, on one or two occasions, we have known them allowed to take honours in the general "Black Spanish" class; but usually they are quite overlooked by poultry judges.

We think it would be well worth while to try the effect of throwing a cross of this breed into its more aristocratic relative. The *hen* should be selected for the cross, of course—not only to avoid the risk of contaminating a whole strain by the experiment, but because it is chiefly size and constitution that are wanted, while the red face must be as speedily as possible "bred out" again. Let a fine Minorca hen, therefore, be put with a good white-faced cock, and her eggs carefully kept apart. When hatched, let one or two of the *pullets* only which show most size and constitution be again reserved, and mated with another good cock of a different family, and so on. We have never seen the experiment tried, but believe a few years of this system would breed good white-faced birds, far superior in size and stamina to any of the existing strains.

WHITE.—This breed should have a red face and white ears, as in the Minorca, which it also resembles in size, shape, and general qualities. The plumage, however, is snow-white, without a single stain. All black fowls occasionally throw white chickens, and no doubt the white breed was thus accidentally originated.

Straw-colour in the cock, or stains of red in the ears, are the most common faults in this variety.

ANDALUSIAN.—This must be considered a truly useful and handsome fowl, being, according to general testimony, the hardiest of all the Spanish breeds. The plumage is slaty blue, in many specimens slightly laced with a darker shade, but the neck hackles and tail feathers are glossy black, and harmonise very richly with the rest. Ears white and face red, as in the Minorca. Unlike other Spanish chickens, these are very hardy, and feather rapidly and well, which gives them a great advantage. This breed appears each year to increase the number of its admirers, and may very probably attain in time to a distinct class of its own.

ANCONAS.—Mottled all over, or what is called "cuckoo" colour, and look rather pretty. In all other points they resemble Minorcas, being, however, of a smaller size.

The so-called "Columbian" fowl is evidently a cross between the Spanish cock and Malay hen, but would be well worth establishing as a distinct breed. The black plumage is of extraordinary lustre, whilst the bird is of great size and hardihood, excellent for the table, and the hen a most prolific layer, the eggs being also probably the very largest known. Even as a cross, such fowls are well worth keeping; and there can be no doubt that a well-established breed combining these qualities would soon become a favourite.

Spanish fowls of any kind are very little subject to roup, at least in any marked or specific form; but suffer exceedingly from cold or wet. Severe frost especially often attacks the comb and wattles, and if the bird in this state be not attended to, it will be disfigured for life. The proper treatment is to rub the affected parts with snow or cold water, exactly as in the human subject, but not on any account to take the frost-bitten bird into a warm room until recovered. The fowls are also very long over their moult, and need special care and nourishing food at this season.

They are also liable to a peculiar disease called "black rot." The symptoms are a blackening of the comb, swelling of the legs and feet, and general wasting of the system. It can only be cured in the earlier stages by frequent doses of castor-oil, to keep up purging; at the same time giving freely strong ale or other stimulants, with warm and nourishing food.

Another singular disease occasionally occurring in this fowl has never, we believe, had any name given to it; but the symptom is the occurrence, in rapid succession, of bladders under the skin, which contain however nothing but air. We believe the cause to be debility: at least, nourishing and stimulating food, pricking each vesicle as it rises, will generally effect a cure.

The merit of Spanish fowls is their production of large white eggs, which are laid in great abundance in moderate weather. They are also of very good quality as table-birds. But they cannot be called good winter-layers, unless with the aid of artificial heat; and their delicacy of constitution is a great drawback to their otherwise many merits. We believe, however, that fanciers have this point much in their own hands; and, even in spite of such a serious fault, wherever large eggs are valued or desired, the Spanish will always be regarded as a most useful and profitable fowl—the Minorca being the best regarded from this point.

As a "fancy" fowl we believe the "white-faced" variety to be the most *profitable* of any, as good stock are always saleable at high prices, and out of a dozen good eggs there are almost invariably a larger proportion of chickens fit for exhibition than can be reckoned upon in any other breed we are acquainted with.

For two or three weeks before exhibition, Spanish fowls should be allowed as much meal as they like to eat. The day before sending off, the legs should be carefully washed, and also the comb, wattles, and face; drying the latter carefully with a soft towel. The face will probably get rather red under this treatment, and if so, the bird must be put for the night in a *warm* room, kept perfectly *dark*, which will make all right again. The hamper should also be carefully lined, that the birds may not take cold, and the top should be high enough to avoid any danger of injury to the combs.

To send fowls of this breed to winter shows in a basket not lined, is in severe weather almost certain death.

CHAPTER XVII.

HAMBURGHS.

UNDER the name of Hamburghs are now collected several varieties of fowls, presenting the general characteristics of rather small size, brilliant rose combs, ending in a spike behind, projecting upwards, blue legs, and beautifully pencilled or spangled plumage. None of the Hamburghs ever show any disposition to sit unless in a state of great freedom, but lay nearly every day all through the year, except during the moulting season, whence they used to be called "Dutch every-day layers."

It is not our province to enter into the question of the origin of the different breeds of Hamburghs. There can be no doubt that the usual classification into simply spangled and pencilled is not sufficient to mark the distinct varieties that exist; but our duty is to take the classes as we find them, and describe them as they are now recognised at the leading shows; paying special attention to the plumage, as exactness of marking is of more importance in this than in almost any other breed. In so doing we are glad to acknowledge the able assistance of Mr. Henry Beldon, of Goitstock, Bingley, Yorkshire, who at present breeds these beautiful varieties more extensively, and takes more prizes, than any one else in the kingdom.

SILVER-PENCILLED.—The size of this exquisite breed is small, but the shape of both cock and hen peculiarly graceful and sprightly. Carriage of the cock very conceited, the tail being borne high, and carried in a graceful arch. The comb in this, as in all the other varieties, to be rather square in front, and well peaked behind, full of spikes, and free from hollow in the centre. Ear-lobe pure white, free from red edging. Legs small and blue.

The head, hackle, back, saddle, breast, and thighs of the cock should be white as driven snow. Tail black, glossed with green, the sickle and side feathers having a narrow white edging the whole length, the more even and sharply defined the better. Wings principally white, but the lower wing-coverts marked with black, showing a narrow indistinct bar across the wing. The secondary quills have also a glossy black spot on the end of each feather, which gives the wing a black edging. The most frequent defect in the cock is a reddish-brown patch on the wing, which is fatal. We believe this fault to occur nearly always in old birds, and remember seeing a cock which had taken thirty-seven prizes moult out thus at last, and so end his career as an exhibition bird. The bar on the wings is difficult to get, and is not imperative; any cock with a nicely edged tail, and quite free from coloured or black markings on any part of the body, ought to stand a fair chance in exhibition, if form and comb be good. As a bird to breed from, however, he would be a failure; as it is impossible to get well-marked pullets except from a cock with a good proportion of black under-colour

The most frequent fault in the hen is a spotted hackle, instead of a pure white. The rest of the body should have each feather distinctly marked, or "pencilled" across with bars of black, free from cloudiness, or, as it is called, "mossing." (See "Feathers," No. 5.) The tail feathers should be pencilled the same as the body; but to get the quill feathers of the wings so is rare, and a hen thus marked is unusually valuable. General form very neat, and appearance remarkably sprightly.

GOLDEN-PENCILLED.—The form of this breed is the same as the preceding variety, and the black markings are generally similar, only grounded upon a rich golden bay colour instead of a pure white. The cock's tail should be black, the sickles and side feathers edged with bronze; but tails bronzed all over are often seen. The bar on the wing is not imperative, or even usual, in this breed. The colour of the cock is always much darker than that of the hens, generally approaching a rich chestnut.

In all pencilled Hamburghs the value chiefly depends on the exactness and definition of the markings, which ought to be a dense black, and the ground colour between quite clear. The silver is slightly the largest breed.

GOLDEN-SPANGLED.—Whilst the markings on pencilled Hamburghs consist of parallel bars across the feathers, the varieties we are now to consider vary fundamentally in having only *one* black mark at the end of each feather, forming the spangle. This black marking varies in shape, and though only one variety is recognised in each colour at poultry exhibitions, it is quite certain that both in gold and silver there are two distinct breeds, distinguished by the shape of the spangle. The best known of the two varieties, and the most often seen, is the breed long known in Lancashire under the name of "mooneys," from the spangles being round, or moon-shaped.

The ground colour of the Golden "Mooney" Hamburghs is a rich golden bay, each of the feathers having a large circle, or moon, of rich black, having a glossy green reflection. (See "Feathers," No. 4.) The hackle should be streaked with greenish black in the middle of the feathers, and edged with gold Tail quite black, even in the hens. All the spangles should be large and regular in shape.

In the cock the upper part of the breast is usually glossy black, but lower down, at least, it ought to be rich bay, and spangled like the hens. The cock of this breed is rather small in proportion to the hens.

The second variety is that known chiefly in Yorkshire as "pheasant fowls," and differs greatly in the plumage. Instead of the spangles being round, as in the "mooneys," they are crescent-shaped (See "Feathers," No. 3), approaching the character of lacing; the marking is also seldom so sharp and definite, being often a little "mossed." In the cock the crescent spangles on the breast run so much up the sides of the feathers as really to become almost a lacing.

The latter variety is the largest, hardiest, and the best layer; but is seldom seen at shows pure-bred. The usual plan appears to be, to show mooney hens along with cocks bred between mooney and pheasant fowls. The reason of this is that the mooney cock has scarcely ever a pure ear-lobe, and generally has a dark breast; and by crossing the two breeds together, cocks are produced with spangled breasts and white ear-lobes, and altogether much larger and showier birds than the pure mooneys. It will be seen, therefore, that to breed birds for exhibition, two distinct lots must be penned up; viz., pure mooneys for the pullets, and mooneys with Yorkshire Pheasants for the cockerels; of course choosing birds for this purpose with the best developed ear-lobes and most evenly spangled breasts. The cross thus obtained may be also used to breed cocks from, but not to breed pullets; although of late even hens have been shown with a taint of the Yorkshire Pheasant in them, as evidenced by their white ear-lobes and larger size. These birds show well under cover, but when seen in full daylight are not to be compared to the true-bred mooney hen in richness of plumage.

We have been careful to explain this at length, because ignorance of it has disgusted many with this truly beautiful breed. Many a "first-prize pen" has been purchased, and the breed afterwards given up in disgust, on account of the cross in the cock not being known or understood, and the pen therefore bred from as in other breeds. We cannot but consider such cross-breeding a great pity; but it is encouraged by the judges, who look mainly for a white deaf-ear; and all we can do therefore is to make the plan of breeding plain to the uninitiated.

SILVER-SPANGLED.—In this class two similar varieties exist. The Lancashire silver "mooney," with large round spangles, resembles the golden, substituting a silvery white ground colour. The outside tail feathers in the hen, however, differ from the golden mooney, being silver white, with only black moons at the tips. The moons on wing covert feathers in both sexes should form two black bars across the wings; the more regular these bars the more valuable the bird.

The silver pheasant-fowl of Yorkshire has smaller spangles, and not so round, without, however, running into the crescent form of the golden pheasant-fowl. The tail is white in both cock and hen, ending in black spangles. The cock's breast has also far less spangling than the mooney breed.

With regard to breeding Silver-spangled Hamburghs for exhibition, the case is still more complicated than in the golden variety, as even the silver-mooney contains two distinct *subvarieties*. The purest strain breeds cocks that are hen-feathered, or marked and feathered exactly like the hen, with the exception that the top feathers of the tail are rather longer. This variety formerly took all the prizes, being larger and much handsomer in marking; but the judges at Birmingham discarded them some years ago, and since then they have nearly died out, being only kept up by a few of the most eminent breeders who know their real merits. There is another and far more common variety of mooney, which breeds cocks with dark tails and reddish ear-lobes, and is probably originally a cross from the higher-bred variety just mentioned with the Yorkshire Pheasant. Be this as it may, cocks for exhibition are usually bred by mating this latter variety again with the Yorkshire Pheasant, the cross producing a cockerel which meets the requirements of the judges, having a full yet clear tail, and pure white ear-lobes; it, however, lacks that depth of colour for which the true-bred mooney is conspicuous. Good pullets can also be bred from the Lancashire dark and full-tailed cock when mated with his *own* hens, but not equal in either colour or size to those bred from the hen-feathered birds; which latter, however, are of little use for breeding cockerels.

We cannot avoid remarking on the folly of these ingenious and yet clumsy proceedings, so opposed to real scientific breeding. The proper plan would have been to adopt as a basis the most perfect variety—the hen-tailed mooney—and by careful selection of breeding stock, to banish that feature when found to be objectionable, which might have been done in a few years, all other merits of marking being retained. Instead of this, we find a system which infallibly disgusts every one ignorant of its mysteries with the whole breed; and whether judges or exhibitors are most to blame for it, it would be hard to decide.

There is also a hen-feathered Golden Hamburgh; and in both colours these birds are very hardy and long-lived. But it is to be noted that Silver-spangled Hamburghs, both Lancashire and Yorkshire, are much better layers than the Golden birds.

It should be noted that many spangled Hamburgh chickens

are at first *pencilled* in the feathers, the true spangling only appearing with the first moult.

BLACK HAMBURGHS.—There is little doubt that this breed has been produced by crossing with the Spanish; the white face often half-apparent, the larger size (cocks often weighing 7 lbs.), and the darker legs, all betray its origin. It is, however, perfectly well established as a distinct variety, and good strains breed quite true to colour and other points.

In the black variety the comb of the cock is considerably larger than in the others we have noted, the wattles also being large and round. Plumage black, spangled, when seen in the light, with dark glossy green. Hens similar in plumage; but in general make rather square and heavy, with short legs, very different from the other varieties.

On the whole, we can most strongly recommend Hamburghs as a profitable breed. Each hen will lay from 200 to 250 eggs in a year, which certainly exceeds the production of any other fowl; and if they are generally small, the consumption of food is comparatively even more so. Though naturally loving a wide range, there is no real difficulty in keeping them in confinement, if cleanliness be attended to. Last year we hatched a brood of eleven, two of which were killed, and all the remainder we have now. Till three months old they had the run of the garden, since which they have been chiefly confined in a shed; but are all in perfect health, and well repay their food. Indeed, more profitable fowls are none; whilst their varieties of barring, pencilling, or spangling, with their elegant shape, form the very perfection of bird beauty, and never fail to excite admiration.

The great difficulty in keeping them arises from their erratic propensities. Small and light, they fly like birds, and even a ten-feet fence will not retain them in a small run. They may, it is true, be kept in a shed; but, if so, the number must be very limited. Where six Brahmas would be kept, four Hamburghs are quite enough, and they must be kept dry and *scrupulously* clean. The pencilled birds are also, most certainly, delicate, being very liable to roup if exposed to cold or wet; they should not, therefore, be hatched before May. The spangled are hardy, and lay larger eggs than the pencilled; but the latter lay rather the most in number. For profit, however, we should recommend the black Hamburgh, on account of the large size of the eggs; and this variety is certainly the most extraordinary egg-producer of all breeds known.

Hamburghs are too small to figure much on the table. They carry, however, from the smallness of the bones, rather more meat than might be expected, and what there is of it is of first-rate quality and flavour.

CHAPTER XVIII.

POLANDS.

UNDER the title of Polands, or Polish fowls, should be collected all varieties which are distinguished by a well-developed crest, or tuft of feathers on the top of the head. This crest invariably proceeds from a remarkable swelling or projection at the top of the skull, which contains a large portion of the brain; and it is worthy of remark, that as the comparative size of this protuberance invariably corresponds with that of the crest springing from it, the best crested chickens can be selected even when first hatched. It is also remarkable that the feathers in the crest of the cock resemble those of his neck-hackles, being long and pointed, whilst those of the hen are shorter and round; and this difference forms the first means of distinguishing the sexes.

The comb of all Polish fowls is likewise peculiar, being of

what is called the two-horned character. This formation is most plainly seen in the Crèvecœurs, where the two horns are very conspicuous. In the breeds more specifically known as Polish, the comb should be almost invisible, but what there is of it will always show the bifurcated formation.

Under the title of Polish fowls, might perhaps be included the Crèvecœurs, Houdans, and Gueldres, if not La Flèche; but we shall, for convenience of reference, describe these crested fowls in a separate chapter on the French breeds, and confine ourselves here to the other tufted varieties, including the recently introduced Sultans.

The following descriptions have been corrected to the latest date by Mr. Henry Beldon, of Bingley, Yorkshire, well known as a prize-taker with these breeds.

WHITE-CRESTED BLACK.—This is the most generally known of all the varieties. The carriage of the cock, as in all Polands, is graceful and bold, with the neck thrown rather back, towards the tail; body short, round, and plump; legs rather short, and in colour either black or leaden blue. There should be no comb, but full wattles of a bright red; ear-lobes a pure white. Plumage black all over the body, with bright reflections on the hackle, saddle, and tail. Crest large, regular, and full, even in the centre, and each feather in a *perfect* bird we suppose of a pure white; but there are *always* a few black feathers in front, and no bird is therefore to be disqualified on that account, though the fewer the better. Weight from five to six pounds.

Hen very compact and plump in form. Plumage a deep rich black. Crest almost globular in shape, and in colour like the cock's. We never yet saw a bird in whose crest there were not a *few* black feathers in front, and we doubt if such were ever bred. Where they do not appear, we believe the crests have always been "trimmed," and in no class does this practice so frequently call for the condemnation of the poultry judge.

SILVER STANGLED POLANDS.

Weight of the hen four to five pounds. This variety is peculiarly delicate and subject to roup.

BLACK-CRESTED WHITE.—There is indisputable evidence that there once existed a breed of Black-crested White Polands; but, unfortunately, it is equally plain that the strain has been totally lost. The last seen appears to have been found by Mr. Brent, in 1854, at St. Omer, and if the breed still exists at all, we believe it will be found either in France or Ireland. Its disappearance is the more to be regretted, as it seems to have been not only the most ornamental, but the largest and most valuable of all the Polish varieties. The hen described by Mr. Brent dwarfed even some Malay hens in the same yard.

We believe the *colour* of this variety may be recovered by breeding from such birds of the kind next mentioned as show any tendency to black in the crest, and carefully selecting the darkest crested chickens. Mr. W. B. Tegetmeier did commence such an experiment, and succeeded perfectly in producing white chickens with black crests, though they always became more or less marked with white in subsequent moults. The attempt was therefore discontinued, though a few years' longer perseverance would undoubtedly have established the strain true to colour, in accordance with the principles laid down in Chapter VIII. But the great comparative size, which all accounts agree belonged to the old breed, we are afraid is for ever lost.

WHITE-CRESTED WHITE.—This breed, and those which follow, differ from the white-crested black Polands not only in greater hardihood, but in having a well-developed beard under the chin, in lieu of wattles. They are large fine birds, and the crest is finer and more perfect than in most other colours. They are also among the best in point of laying. The plumage needs no description, being pure white throughout.

SILVER SPANGLED.—In this variety the ground colour of the plumage is a silver white, with well-defined moon-shaped black spangles. (See "Feathers," No. 4). In the cock, the hackle feathers are white, edged and tipped with black; in the hen each hackle feather should have a spangle on the end. Tail feathers clear white, with a large spangle on the ends. The spangling on the wing coverts should be large and regular in both sexes, so as to form *two* well-defined bars across each wing. The proper spangling of the breast is very important. Many cocks are nearly black on the upper part, which is a great fault.

The crest feathers are black at the base and tip, with white between. Crest to be full and regular, showing no vacancy in the centre. A few white feathers usually appear after the second moult in the very best bred hens, and in old birds are not a disqualification, though certainly a fault.

Ear-lobes small and white; wattles none, being replaced by a black or spangled beard.

The size of this breed is very good, weight of the cock six to seven and-a-half pounds; hens four to five and-a-half pounds.

Besides the moon-shaped spangling, birds are shown with laced feathers, that is, with an edging of black on the outline of the feathers, but thicker at the end. This marking when perfect is of exquisite beauty, and appears at present to secure most of the prizes. The hens have clear white tails, laced and spangled with black; but the cocks have generally dark tails, which takes much from their beauty. A few have, however, been shown with clear tails like the hens, but appear to find no favour with the judges, who seem to prefer the darker-looking birds, inferior in beauty as they nevertheless certainly are. From such laced birds were derived the celebrated Sebright Bantams.

GOLDEN-SPANGLED.—This breed is similar to the preceding in the black markings, substitutiug a rich golden ground for the silver white. The tail of the cock, however; is dark bay, the sickles being tipped with black, and the side feathers edged evenly with the same colour. Like the preceding variety, golden-spangled Polands are also very often shown with the markings in the form of a lacing, and such are just now most popular.

BUFF or CHAMOIS POLANDS are a recent introduction. This breed resembles the golden-spangled in the colour of the ground, but the spangles present the anomaly of being *white* instead of black. They were first produced, there can be no doubt, by crossing the golden-spangled with white birds, and even yet they do not appear to have been thoroughly established or bred exactly true to colour. The appearance is very pretty, and the variety will no doubt become a favourite.

Blue, grey, and cuckoo or speckled Polands are also occasionally shown, but are evidently either accidental occurrences, or the result of cross-breeding, and cannot be recommended even to the fancier.

All the bearded Polands are rather liable to grow up "hump-backed," or "lob-sided" in the body. Of course either defect is a fatal disqualification.

SULTANS.—This breed was introduced by the well-known Miss E. Watts, of Hampstead, and is a very ornamental bird, differing greatly in appearance from any of the varieties hitherto named. In size they are rather small, the cocks weighing only from four to five pounds. They make most exquisite pets, being very tame, but at the same time brisk and lively; and their quaint little ways never fail to afford much amusement. They appear well adapted to confinement.

The plumage is pure white, crest included, in which they therefore resemble the white Polands. They differ, however, very greatly in appearance. Their legs are very short, and feathered to the toes; the thighs being also abundantly furnished, and vulture-hocked. They are likewise amply muffed and whiskered round the throat, and the tail of the cock is remarkably full and flowing. The crest differs from that of most other Polands, being more erect, and not hiding the eyes.

The comb consists of two small spikes in front of the crest. The legs are also white instead of blue, and the foot has a fifth toe, like the Dorking fowl. The adult birds appear hardy.

There is a breed known as Ptarmigans, which is evidently a degenerate descendant from some former importation of Sultans.

Some special precautions are necessary in rearing Polish chickens. The prominence in the skull, which supports the crest, is never completely covered with bone, and is peculiarly sensitive to injury. On this account Cochins, or other large heavy hens, should never be employed as mothers. A game hen will be the best. The young also fledge early and rapidly, and usually suffer severely in the process; they therefore require an ample allowance of the most stimulating food, such as hemp-seed, meat, and bread steeped in ale; and, above all, they *must* be kept dry.

Polands have certainly solid merits. They improve in appearance, at least up to the third year. In a favourable locality they are most prolific layers, never wanting to sit, and the flesh is remarkably good. They appear also peculiarly susceptible of attachment to their feeders. And lastly, they suffer remarkably little in appearance or condition from exhibition.

Their great fault is a peculiar tendency to cold and roup—the white-crested black variety being the most delicate of all. The dense crest becomes during a shower saturated with water, and the fowls are thus attacked in the most vital part. No birds are so affected by bad weather. In exposed or damp situations they will die off like rotting sheep, and it is hopeless to expect any return. They can only be kept successfully in warm, genial situations, on well-drained ground, with a chalk or sand sub-soil, and with ample shelter to which they can resort during showers. In such circumstances they will do well, and repay the owners by an ample supply of eggs.

Mr. Hewitt cautions Polish breeders against attempting to seize their birds suddenly. The crest so obscures their vision that they are taken by surprise, and frequently so terrified as to die in the hand. They should, therefore, always be first spoken to, or otherwise made aware of their owner's approach.

CHAPTER XIX.

FRENCH BREEDS.

SINCE the fancy for poultry breeding spread in some degree to our Gallic neighbours, several remarkable breeds of fowls have been introduced into England from France, which it will be convenient to describe in one chapter. They all deserve especially the careful attention of the mercantile poultry breeder, possessing as they do in a very high degree the important points of great weight and excellent quality of flesh, with a remarkably small proportion of bones and offal. These characteristics our neighbours have assiduously cultivated with most marked success, and we cannot avoid remarking yet again on the results which *might* have been produced in this country had more attention been paid to them here, instead of laying almost exclusive stress upon colour and other fancy points.

Most of the French breeds have more or less crest, which naturally places this chapter next to that on the Polish fowls. It is remarkable also that they all agree in being non-sitters, or at least incubate but very rarely.

CRÈVECŒURS.—This breed has been the longest known in England, and is the one most preferred in France for the

152 DIFFERENT BREEDS OF FOWLS.

CRÈVECŒUR.

quantity and quality of its flesh. The full-grown cock will not unfrequently weigh 10 pounds, but 7½ to 8 pounds is a good average.

In form the Crève is very full and compact, and the legs are exceedingly short, especially in the hens, which appear almost as if they were creeping about on the ground. In accordance with this conformation, their motions are very quiet and deliberate, and they appear the most contented in confinement of any fowls we know. They do not sit, or very rarely, and are tolerable layers of very large white eggs.

The comb is in the form of two well-developed horns, surmounted by a large black crest, and giving the bird a decidedly "diabolical" appearance. Wattles full, and, like the comb, a very dark red. The throat is also furnished with ample whiskers and beard.

Plumage mostly black, but in the largest and finest birds not unfrequently mixed with gold or straw on the hackle and saddle. Which is to be preferred will depend upon circumstances. Judges at exhibitions always prefer a pure black all over; and if the object be to obtain prizes, such birds must of course be selected both for breeding and show purposes; at the same time we should fail in our duty were we not distinctly to record our opinion that such a choice is most unfortunate, as the golden-plumaged birds are generally by far the largest and finest specimens. It should be remembered that the French have mainly brought these breeds to perfection by seeking first the *useful* qualities, and if our "feather-breeding" propensities be applied to them, we much fear that uniformity will only be attained at the price of the deterioration of the strain in size and real value.

The merits of the Crève consist in its edible qualities, early maturity, the facility with which it can be both kept and reared in confinement, and the fine large size of its eggs. The hen is, however, only a moderate layer, and the eggs are often sterile, while the breed is rather delicate in this country, being subject to roup, gapes, and throat diseases. This delicacy of constitution appears to improve somewhat as the fowls get acclimatised, and we should, therefore, recommend good English-bred rather than imported birds. Altogether, we do not recommend the Crève as a good breed for general domestic purposes; but it is certainly a splendid fowl for either table or market, and as such, especially on a large scale, in favourable localities, well repay the breeder.

Our engraving was drawn in France from remarkably good and perfect specimens.

LA FLÈCHE.—In appearance this breed resembles the Spanish, from which we believe it to have been at least partly derived. It exceeds that breed, however, in size, the cock often weighing from eight to even ten pounds. Both sexes have a large, long body, standing on long and powerful legs, and always weighing more than it appears, on account of the dense and close-fitting plumage. The legs are slate-colour, turning with age to a leaden grey. The plumage resembles the Spanish, being a dense black with green reflections.

The look of the head is peculiar, the comb being not only two-horned, much like the Crèvecœur, near the top of the head, but also appearing in the form of two little studs or points just in front of the nostrils. The head used to be surmounted by a rudimentary black crest, but English fanciers very soon bred this out, and the presence of crest is now considered a disqualification at all good shows. The wattles are very long and pendulous, of a brilliant red colour, like the comb. The ear-lobes are dead white, like the Spanish, and exceedingly developed, meeting under the neck in good specimens. In fact, no breed could show stronger traces of its Spanish origin.

LA FLÈCHE.

The appearance of the La Flèche fowl is very bold and intelligent, and its habits active and lively; at the same time it does not appear to thrive well in our climate. The hen is an excellent layer of very large white eggs, and does not sit. The

La Flèche Cockerel.

flesh is excellent, and the fine white transparent skin makes a very favourable appearance on the table, which is only marred by the dark legs. The breed is, however, very delicate, and does not lay well in winter, except in favourable circumstances. Altogether, it is decidedly less suitable than the preceding for domestic purposes, but still most valuable as a table fowl. As an egg producer, it is as nearly as possible similar to the Spanish, not only in the size and number of the eggs, but the seasons and circumstances in which they may be expected. In

La Flèche Pullet.

juiciness and flavour the flesh approaches nearer to that of the Game Fowl than any other breed we know.

The cocks suffer much from leg weakness and disease of the knee-joint, and do not bear the fatigue and excitement of exhibition so well as most fowls. They require, therefore, special care, and the moderate use of stimulants.

HOUDANS.—This fowl in many respects resembles the Dorking, and Dorking blood has evidently assisted in its formation. We believe that a cross between the latter and a white Poland would not be very wide of the mark. Houdans have the size deep compact body, short legs, and fifth toe of the Dorking,

Houdan Cock.

which in form they closely resemble, but with much less offal and smaller bones. The plumage varies considerably, but is most usually white, with large black spangles, the size of a shilling in many specimens. We should certainly like to see the spangling reduced in size of the markings, but sincerely hope this will not be sought at the expense of weight, in which the Houdan is pre-eminent among the French breeds. We feel certain that by breeding for this more useful quality the fowl may be reared to a greater weight than even the coloured Dorking; we have ourselves seen hens which weighed ten pounds, but such a size is not common, and very small speci-

Houdan Hen.

mens are more often seen at exhibitions than of the other French varieties.

The head should be surmounted by a good Polish crest of black and white feathers. The wattles are pendent and well developed, and the comb is the most peculiar in formation of all the French breeds, resembling, as has been said, the two leaves of a book opened, with a long strawberry in the centre; in the hen it should be very small and rudimentary.

Imported Houdans frequently want the fifth toe, evidently

BREDA OR GUELDRES. 159

BREDA.

derived from the Dorking; and it might at this early period be easily bred out. We cannot but express our regret this should not be done, regarding it, as we do, not only as an eyesore, but in our opinion more or less connected with the diseased foot of the Dorking fowl. The aim of fanciers, however, seems to be to ensure the additional toe by careful selection, and in a few years it will be established as an indelible feature.

With respect to the merits of Houdans, we have no hesitation in pronouncing them one of the most valuable breeds ever introduced into this country; and in this judgment we are fully corroborated by Mr. F. H. Schröder, of the National Poultry Company, who expressed to us his strong opinion that in general usefulness Houdans surpassed all the French varieties, to which the company devote their principal attention. We have in this breed the size, form, and quality of the Dorking, with earlier maturity. The hen is a most prolific layer of good-sized eggs, which will almost invariably be found *fertile*—a point the Dorking is very deficient in, as all prize breeders know to their cost. The chickens feather very rapidly and early, but are nevertheless *exceedingly* hardy, perhaps more so than any except Cochins or Brahmas, and are therefore easily reared with little loss. They are emphatically the fowl for a farmer, and will yield an ample profit on good feeding, both in eggs and flesh.

Almost their only drawback is their refusal to incubate. Many, however, will consider this an advantage. The bird will bear a moderate amount of confinement well, but in this respect is not quite equal to the Crèvecœur.

BREDA OR GUELDRES.—This fowl is of exceedingly well-proportioned shape, with a wide, full, prominent breast. The head carries a small top-knot, and surmounts a rather short, thick neck. The comb is very peculiar, being hollowed or depressed instead of projecting, which gives to the head a most singular expression. Cheeks and ear-lobes red; wattles ditto, and in the cock very long and pendulous.

The thighs are well furnished and vulture-hocked, and the shanks of the legs feathered to the toes, though not very heavily. The plumage varies, black, white, and cuckoo or mottled, being most seen. The cuckoo-coloured are known exclusively by the name of "Gueldres," and the black bear chiefly the name of Bredas; but it is much to be desired that one name should be given to the whole class, with simply a prefix to denote the colour. We prefer ourselves the black variety, the plumage of which is beautifully deep and rich in tone, with a bronze lustre; but Mr. F. Schröder, who thinks highly of the breed, prefers the cuckoo or Gueldres fowl. This is quite matter of fancy, all the colours being alike in economic qualities.

The flesh is excellent and tolerably plentiful, very large cocks weighing as much as eight or nine pounds. They are very good layers, and the eggs are large; like the other French breeds, the hens do not sit. The chickens are hardy, and the breed is decidedly useful and well adapted to the English climate.

Our illustration is drawn from a very good pair of the Black or Breda variety.

LA BRESSE.—This fowl is hardy and large, but we cannot, at present at least, consider it as a distinct or established breed. The birds are *all* colours without distinction, presenting exactly the appearance of very large and fine barn-door or cross-bred fowls; and we believe that it is, in fact, no breed, but a mixture of fine specimens of different races. A few years' breeding in England will decide this, and may possibly produce some uniformity in colour. At present we can only say that the shape and size should be as nearly as possible that of the Grey Dorking, while the colour may be anything. Of course, in an exhibition pen the two hens must match. Mr. Schröder commended the La Bresse fowl to us as pre-eminent for its early maturity and fattening qualities, and we should consider it a valuable addition to our farm-yard stock. By the fancier it will be little prized.

It will be seen that the French breeds are eminently table fowls; and it is worthy of remark that by breeding for edible qualities, without paying over-much attention to feather or other fancy points, our neighbours have succeeded in producing birds far superior to any English breed—we will not say in quality, so long as Game and Dorking are left us—but in smallness of bone and offal. We should hope that the lesson may not be lost upon our breeders, and that poultry committees may be led to afford somewhat more encouragement than they have hitherto done to the cultivation of size and general proportion, with a view to the table, as distinguished from mere artificial or fancy qualities.

Of all the French breeds we should ourselves give the first rank to the Houdan, on account of its great hardihood and plentiful production of eggs. Next in value we would place Gueldres and La Flèche. The Crèves, beautifully heavy birds as they are, we consider too delicate in our climate ever to become a *general* favourite. Others, however, would place them first; and as we have endeavoured to state fairly all points bearing on the subject, we must leave the reader to form his own judgment and make his own comparison. One thing is certain, that all these fowls are composite—are artificially created; and it would be well if a little more enterprising experiment in this direction were made by English breeders.

CHAPTER XX.

BANTAMS.

THERE is not the slightest reason for supposing that any of the diminutive fowls known as Bantams are descended from an original wild stock. They are in many cases the exact counterparts of ordinary domestic breeds, carefully dwarfed

BLACK AND SEBRIGHT BANTAMS.

and perfected by the art of man; and even where this is not so, the process by which they were produced is occasionally on record. They are, in fact, more than any other class, "artificial fowls," and their attractiveness consists rather in their beauty than in any economic value. We can only enumerate and give descriptions of the principal varieties, as drawn up under the able supervision of the Rev. G. S. Cruwys, of Tiverton, long celebrated for his success as an exhibitor and breeder of these beautiful birds.

SEBRIGHTS.—Cock not to exceed twenty, and hen sixteen ounces. For exhibition still less is preferable, but not for breeding. Carriage of the cock, the most conceited it is possible to conceive of; head thrown back till it touches the nearly upright tail; wings drooping halfway down the legs; motions restless and lively, always strutting about as if seeking for antagonists. The bird is, in fact, "game to the backbone," and will attack the largest fowl with the utmost impudence.

Plumage close and compact, and *every feather* laced with black all round the edge. The shoulder and tail coverts are the parts most likely to be faulty in this; but in first-class birds every single feather must be properly edged right up to the head. This part usually appears darker from the smaller size of the feathers; but the nearer the head is to the rest of the body in colour the better. The only exceptions allowable in the lacing are on the primary quills or flight-feathers of the wings, which should have a clear ground, and be only tipped with black. The tail feathers ought to be laced, and in the hen must be so; but in the cock this is rather rare. In his case a clear ground colour throughout, nicely tipped with black, may be allowed to pass instead.

The cock must be perfectly *hen-feathered* throughout, his tail not only square and straight, without sickles, but the neck and saddle-hackles resembling those of the hen. Mr. Hewitt, however, a most eminent authority on this breed, remarks that while this is imperative for exhibition, he has always found such cocks nearly or quite sterile, probably in consequence of the long interbreeding necessary to maintain the strain in perfection. He recommends, therefore, that a cock for breeding should show a moderate approach to sickle-feathering, when the eggs will become productive.

The comb should be a perfect rose, with a neat spike behind, pointing rather upward, free from any depression, and rather livid in colour. Face round the eye rather dark. Eye itself a sparkling dark red. The ear is supposed to be white, but Mr. Hewitt remarks that he never found it so without a great falling off in the lacing of the plumage, and a bluish tinge is as near an approach to it as can be safely obtained. Bill slate-coloured; legs blue and clean.

There are two varieties. In the gold-laced the ground colour is a rich golden yellow. In the silver-laced, a pure white. In both cases the ground must be perfectly clear and unsullied, varied only by the clear black line round each feather, which constitutes the lacing. (*See* plate of "Feathers," No. 2.) Lately the Silver Sebrights have shown a decided golden tinge, which greatly mars their beauty, and which may have arisen from an opinion frequently expressed by a well-known breeder, that the clearest birds were bred from a cross between the gold and silver-laced. If this be the case, the sooner such an opinion is exploded the better, as it has already half ruined the beautiful silver breed.

With respect to the breeding of Sebrights, Mr. Hewitt makes two further remarks. First, that although at three years old the birds become more or less grizzled with white, and therefore greatly deteriorated for exhibition, the stock then produced from them is frequently far superior; and secondly, strange as the fact may seem, that better marked birds are usually obtained by mating a heavily-laced cock with a hen scarcely sufficiently marked, than when both parents are perfect in their plumage.

GAME.—In Game Bantams the plumage is precisely similar to the corresponding varieties of the Game fowl, from which they were undoubtedly obtained by long interbreeding, and continually selecting the smallest specimens, occasionally, perhaps, crossing with a Bantam to expedite the process. The carriage and form must also be similar, and the drooping wing, so common in other Bantams, would infallibly disqualify a pen of Game.

In courage and "bottom" Game Bantams are not behind their larger relatives. In constitution they are the hardiest of all Bantam breeds.

In weight the cock must not exceed one and a half pounds, or the hen twenty ounces.

BLACK.—This is at present one of the most popular Bantam classes. The plumage is a uniform black, with no trace of rust, or any other colour, and, in the cock, with a bright lustre like that of the Spanish fowl. Tail of the cock full and well arched; legs short, dark blue or black in colour, and perfectly clean. Comb a bright red rose. Ear-lobes white; face red, in the latter points resembling the Minorca fowl. Cock not to exceed twenty, hen eighteen ounces.

Black feather-legged Bantams have now and then been shown, but never yet established a footing. Fashion changes, however; and novelties being now much sought after, we are inclined to believe that a good feather-legged black-breed would speedily become a favourite.

WHITE.—This breed should be as small as possible, never exceeding two pounds per pair. Except that the legs are white and delicate, all other points are similar to the Black Bantam, changing the colour of the plumage from black to a spotless white. It should, however, be remembered that while the white ear-lobe is required by *most* judges, as in the black variety, there are some who prefer a *red*, and this latter we must express our own decided opinion is much the smartest looking, and harmonises better with the white plumage. The most usual fault is a yellowish colour in the cock's saddle. A single comb is, of course, fatal.

A very pretty feather-legged White Bantam is not unfrequently seen, and, though long neglected, appears to be coming into fashion again. They are usually rather too large, and attention will have to be paid to this particular if the breed is to become popular.

NANKIN.—This is one of the old breeds of Bantams, and at one time nearly disappeared, but attempts have been recently made to re-introduce it. The ground colour is a pale orange yellow, usually with a little pencilling on the hackle. The best tail, to our fancy, is a pure black, with the coverts slightly bronzed. The comb is rose; and the dark legs should be perfectly clean.

PEKIN OR COCHIN BANTAMS.—This most remarkable of all Bantam breeds has only been introduced a few years, the original progenitors having been stolen from the Summer Palace at Pekin during the Chinese war. They were first shown in 1863. They exactly resemble Buff Cochins in colour and form, possessing the feather-leg, abundant fluff, and all the other characteristics of the parent breed in full perfection, and presenting a most singular appearance. They are not yet common, and the interbreeding necessitated by only one original stock existing, has caused much sterility and constitutional weakness. The strongest birds have been bred by crossing with other feather-legged Bantams to introduce fresh blood, and then breeding back to the pure strain. Pekin Bantams are very tame, and make excellent pets.

JAPANESE.—This is the only addition to our poultry-yards yet imported from Japan, though we should hope yet to receive from that country some accessions also to our larger kinds. The Japanese Bantam is very short-legged, and differs from most of the older varieties in having a very large *single* comb. The colour varies. They are often shown mottled or cuckoo-coloured, but what we like best is a pure white body with glossy, jet-black tail.

Bantam chickens require a little more animal food than other fowls, and, for a week or two, rather extra care to keep them dry. After that they are reared as easily as other fowls, and should indeed be rather *scantily* fed to keep down the size. The hens are good mothers, and are often employed to rear small game; and are not bad layers, if the eggs were only larger. We believe them, however, to produce quite as much for their food as ordinary breeds. But their chief use is in the garden, where they eat many slugs and insects, with very little damage. On this account they may be usefully and profitably kept where a separate poultry-yard is found impracticable. We should prefer the Game variety, as being hardiest; and, being good foragers, five or six of these may be kept in a garden for almost nothing, requiring only a house two feet square to roost and lay in.

Bantam eggs are the very thing to tempt the appetite of an invalid, and are just nicely cooked by pouring boiling water over them upon the breakfast-table.

CHAPTER XXI.

THE "VARIOUS" CLASS.

UNDER this heading we propose to describe, shortly, the principal breeds of poultry which usually appear in the class "for any other variety" at our shows, but have never established their claim to a special class of their own. For the most part these breeds have little economic value, but are too well marked in their characteristics to be entirely passed by.

DUMPIES, OR CREEPERS.—This is probably the most useful variety of any mentioned in this chapter, and under various names, such as Go Laighs and Bakies, has long been known and valued in Scotland, though never popular in this couutry. The principal characteristic is the extreme shortness of the shank, or leg bone, which should not exceed two inches from the hock joint to the ground. In other respects they most resemble Dorkings, lacking, however, the fifth toe, and being far more hardy than that variety. The hens are good layers of rather large eggs, and as mothers cannot be surpassed. The plumage is generally an irregular speckle, and it is difficult to get them any uniform colour. The cock should weigh six or seven and the hen five or six pounds.

Dumpies certainly deserve to be better known. They have no particular faults, and, combining as they do very fair laying with great hardiness and first-class edible qualities, they must be considered decidedly profitable fowls. Their extreme shortness of leg also points out their value as a cross to correct the "stiltiness" of some of our larger breeds, whilst the whiteness and quality of the flesh would be improved at the same time.

REDCAPS.—This must also be regarded as a profitable fowl. It is a kind of Golden-spangled Hamburgh, with the difference of being almost as large as a Dorking, and having the rose comb most prodigiously developed, that of the cock being often three inches across, and too heavy to stand upright. They cannot therefore be regarded as ornamental, which is probably the reason they do not meet with much general approval, being moreover often dark on the breast, and far inferior to the Hamburghs in beauty of marking. But they are enormous layers, not to be surpassed by *any* variety; and, with the advantage of a large, plump body, we must pronounce them to be in economic value equal to any we know. They are hardy and easily reared, but not often met with except in Yorkshire, whence they should be procured if a good stock is desired.

SILKY, OR NEGRO FOWLS.—This breed possesses two distinct peculiarities. The webs of the feathers have no adhesion, and the plumage is therefore "silky," or consisting of a number of single filaments, which makes the bird appear much larger than it really is, the actual weight of the cock being generally under three pounds, and of the hen about two pounds. The colour is usually pure white, but other colours are occasionally seen. The second peculiarity is the dark tint of the bones and skin, from which the name of "negro" fowls is derived. The skin is of a very dark violet colour, approaching to black, even the comb and wattles being a dull dark purple. The bones also are covered with a nearly black membrane, which makes the fowl anything but pleasant to look at upon the table; but if the natural

Silky Fowls.

repugnance to this can be overcome, the meat itself is white, and very good eating, indeed superior to that of most other breeds.

The plumage is often so excessively developed as to give the birds a most grotesque appearance. Our illustration is not in the least exaggerated, and is a good representation of many specimens of the breed.

The comb varies in shape; but a Malay comb is best. There is generally a small crest on the top of the head. The legs are mostly well feathered to the ground, and often have five toes; but neither point is universal.

The sole value of the Silky Fowl is as a mother to Bantam, or other small and delicate chickens, such as pheasants or partridges. For such purposes they are unequalled, the loose long plumage affording the most perfect shelter possible. They are, of course, peculiarly susceptible to cold or wet, and have no other value than that stated, except from their singular and not unornamental appearance.

The EMU, or SILKY COCHIN is an occasional sport from the ordinary Cochin fowl. The plumage resembles that of the preceding variety; but in every other point the fowl is a true-bred Cochin. The loose feathering being no real protection from wet, this breed, like the other, is very delicate in our climate.

FRIZZLED FOWLS present a most remarkable appearance, every feather in good specimens being curved, or turned back from the body, so as to show a portion of the under side, like the curved feathers in the tail of a common drake. The colour of the plumage is generally white, with single combs; but double combs and various colours are also seen.

Frizzled fowls are, as might be supposed, exceedingly delicate, and most uncertain layers. The flesh is also inferior, and they have therefore no economic value, whilst they cannot even be termed ornamental. Their only recommendation is their singularity, in which certainly it would be very hard to surpass them.

RUMPLESS FOWLS are of various colours, the only essential characteristic being the absolute want of a tail, or of any approach to one. It is, indeed, exceedingly difficult to breed any particular colour, as few persons have interest in the breed sufficient to persevere long enough for securing uniformity. The handsomest are white: black also looks well; but speckled are

BARN DOOR.

most common. The size also varies very much, ranging from three to seven pounds each.

In this variety not only are the tail feathers absent, but the caudal vertebræ are either wanting altogether, or only rudimentary. The hens are usually very fair layers, sitters, and mothers, and of average quality for the table; the eggs, however, are very apt to prove sterile. On the whole, the breed has few decided faults, and is hardy; the only reason, we suppose, that it is not generally bred, being that birds certainly look handsomer *with* a tail than without one.

RUSSIANS.—This breed is mostly kept in Scotland. It is decidedly a "rough-looking" fowl, being ornamented with tufts of feathers on each jaw, and an abundant beard under the chin. The colour is generally white or buff; but black is also seen, with mixed colours also; the most valuable are spangled like Hamburghs. On the whole, this is a good useful fowl, the flesh being satisfactory, whilst the hens are good layers and rather small eaters. The breed is hardy.

Other fowls are occasionally shown, but do not require special notice, and we believe are very often mere accidental offshoots, or crosses, from well-known breeds. Some few, such as Rangoons and Chittagongs, are evidently chiefly Malay in their parentage; but fowls are constantly shown in the class for "any other distinct varieties" which would defy any attempt to describe their origin. When the parentage is evident, the principal value of such specimens is to show the effect of *crossing*, in which respect they are often useful. This part of poultry-breeding is too much neglected. By it in a great measure has all other agricultural stock been brought to its present perfection; and when steady effort shall be made to *combine* the qualities of some of our best varieties, establishing the strain afterwards by careful selection, we believe we shall have a breed of fowls which in size, prolificacy, and edible qualities united, will surpass any kind hitherto known.

AMERICAN FOWLS

SINCE the early editions of this Work were published, several breeds have been introduced into England from America. They are all of the "useful" type, rather than the ornamental; but one of them, at least, is rapidly becoming a favourite, and an "American" class at shows is almost always well filled.

DOMINIQUES are probably one of the oldest varieties, being only a fixed type of those "Cuckoo" fowls which have always been such favourites with hen-wives. They resemble, in fact, the Cuckoo-coloured fowls known as Scotch Greys, with the exception of having rose-combs and yellow legs. They are plump and tender on the table, and capital layers, and might make a valuable cross for the Cuckoo Dorking. They are, in our opinion, never likely to make a popular variety for showing; but we know several who have kept them for their merely useful qualities, and in no one case has there been disappointment in this respect.

LEGHORNS much resemble in body a small Spanish fowl, having the same large combs (upright in the cocks and falling over in the hens). They have, however, red faces like Minorcas, with a white deaf-ear much smaller than those of Spanish, and all the varieties have bright yellow legs. There are White, Cuckoo-coloured or Dominique, and Brown breeds, the latter the same colour in both sexes as our Black-breasted Red Games. All are hardy, and the *most astonishing layers* we know, averaging over 200 eggs per annum in many cases. The colours are about alike in economic value : in America the

Brown Leghorns are much the most popular, and realise high prices; but in England, so far, the White variety has met with by far the most cordial reception. The Leghorn was undoubtedly imported from the Mediterranean, and its extreme hardiness is therefore remarkable. The only mishap to which it is subject is the large comb becoming frost-bitten, which many Americans guard against by dubbing. It is a non-sitting breed, and lays the largest egg for its size of any fowl we know. Its economic value may be described as that of a Hamburgh, but laying a much larger egg, besides being—what the Hamburgh is not—well adapted for confinement.

PLYMOUTH ROCKS are apparently only a cross between Dominiques and Cochins. They nearly resemble Cuckoo Cochins in all but having clean legs; and we have seen traces of feather which make this distinction doubtful. We cannot speak very highly of this variety as yet, as it appears to breed with little certainty, and to be far from a good layer. Even in America this breed has never become extensively popular, and can only be recommended to those who desire a large bird of the Asiatic type without the accompaniment of leg-feather.

TURKEYS, ORNAMENTAL POULTRY, AND WATERFOWL.

CHAPTER XXII.

TURKEYS. GUINEA-FOWL. PEA-FOWL.

TURKEYS.—The most opposite opinions have been expressed by different breeders as to whether or not the rearing of turkeys in England can be made profitable; and the *general* judgment, we are bound to say, seems to be that they can barely be made to repay the cost of their food. There are not wanting, however, those who from their own experience maintain the contrary; and we believe that where the balance-sheet is unsatisfactory, the cause will generally be found in heavy losses from want of care. The usual mortality in turkey chicks is tremendous, and quite sufficient to eat up any possible amount of profit; but there are many persons who for years have reared *every chick;* and, under these circumstances, they will yield a very fair return.

Without depending upon any one single breeder, we have taken much pains to gather, from the best authorities, the essentials of such successful management; and wherever our directions shall be found to differ from others previously published, the reader may rely with confidence that the treatment given is such as has been thoroughly tested and proved to give the best results.

The first main point to remember is, that for about the

first six weeks or two months the turkey chicks are *excessively* delicate, and that the very slightest shower, even in warm weather, will often carry off half of a large brood. When about two months old, however, the red naked protuberances about the neck and throat begin to appear, and as soon as these are fairly developed, the *chicks* become *poults*, and are soon hardier than any other fowl, braving any weather with impunity.

It is therefore well worth while, and absolutely necessary to pecuniary success, to provide special shelter for the young broods during the critical period, ordinary poultry accommodation being insufficient. Even damp ground is so fatal that a boarded floor is advisable. When any number are to be reared—and we certainly cannot recommend for profit the rearing of turkeys on a *small* scale—we should advise the erection of either a very spacious shed, floored with plank, or a large building of one storey high, to be devoted entirely during the season to the turkey stock. Of course, by a building we mean a mere shell of four bare walls, well roofed, and well lighted. With shelter of this kind there hardly need be a chick lost, except from accident.

It has been stated by many that the number of hens allowed to a turkey cock may be unlimited; and it certainly does appear indisputable that one visit to the cock is sufficient to render fertile all the eggs laid by a turkey hen. The best breeders, however, affirm that as the number of hens allowed to one bird approaches a score, the chicks show falling off in constitution; and the number ought therefore to be limited to twelve or fifteen—quite enough brood stock for even a large establishment. The turkey cock may be used for breeding at two years old, and the hen at twelve months, but are not in their prime till a year older. They will be first-class breeding stock, as a rule, for at least two years later, and many cocks in particular will breed splendid chickens for considerably longer;

a *good* bird should not therefore be discarded till his progeny show symptoms of degeneracy.

The size of the hens is of special importance, much more than that of the cock, in whom good shape, strength, and spirit are of more value, if combined with a fair good size.

The turkey-hen generally lays about eighteen eggs—sometimes only ten or a dozen, and when each egg has been taken away when laid, it may be more. We once heard of ninety eggs being laid by a turkey-hen, but can scarcely credit such a statement. A very good plan is to give a turkey's first seven eggs to a common hen—quite as many as she can cover—when there will be generally just about enough laid subsequently to be hatched by the turkey herself. The best time to hatch the chicks out is in the months of May and June, or even July; and all eggs set should be marked, as the turkey often lays several after commencing incubation.

In a state of nature, the turkey-cock is constantly seeking to destroy both the eggs and chickens, which the female as sedulously endeavours to conceal from him. There is generally more or less of the same disposition when domesticated, and, when it appears, it must be carefully provided against; but the behaviour of very many cocks is quite unexceptionable; and as such a quiet disposition saves a great deal of trouble, it is always worth while to ascertain the character of the cock of the year in this respect. If he be friendly to the chicks and sitting hens, he may be left at large; if otherwise, he must be kep away.

The turkey-hen is very prudish, but gives scarcely an trouble while sitting. She sits so constantly that it is needful to remove her daily from her nest to feed, or she would absolutely starve. Nevertheless, when absent she is apt to be forgetful, and therefore, if allowed to range at liberty, care should be taken that she returns in time—twenty minutes. A better plan, however, is to let her have her liberty only in a confined run of grass. Besides her daily feed, a water vessel and some soft food should be always within her reach. No one must visit the hatching-house but the regular attendant, or the hens will get startled, and probably break many eggs, which easily happens, from the great weight of the birds.

Many have alleged that the turkey sits thirty-one days. This is an error. The chicks break the shell from the twenty-sixth to the twenty-ninth day, scarcely ever later. The day but one before the hatching is expected, the hen should be plentifully fed, the nest cleaned of any dung or feathers during her absence, and an ample supply of food and water placed where she can reach it, as she *must not again be disturbed* till the chicks are out. In dry weather, if the nest be in a dry place, the eggs will have been daily sprinkled as described in Chapter IV. With these precautions, there will rarely fail to be a good hatch.

The egg-shells may be cleared away after hatching has proceeded some hours, but the chicks should *never be taken away from the hen*, and never be *forced to eat*. The latter practice is very general, as turkey chicks are very stupid, and do not seem to know how to peck. But a much better plan is to put two ordinary hen's eggs under the turkey, five or six days after she began to sit, which will then hatch about the same time as her own, and the little chickens will teach the young turkeys, quite soon enough, what they should do. Water or milk may be given, however, by dipping the tips of the finger or a camel-hair pencil in the fluid, and applying it to the end of their beaks.

The usual feeding is oatmeal and bread-crumbs, mixed with boiled nettles. Such food is not good, as turkey chickens for a few weeks have a great tendency to diarrhœa, which the oatmeal rather increases, and the result is a weakening of the system, and frequently many deaths. The very best feeding at first—say for a week—is hard-boiled eggs, chopped small, mixed with *nothing* but minced *dandelion*. With regard to the choice of this herb, Mr. Trotter—who was the first to *study* turkey treatment rationally—and after him many others, have observed that, when at liberty, the young birds invariably choose the dandelion before all other green food, and it probably serves to keep the bowels in proper order. When dandelions cannot be obtained—and it is well worth while to *grow* them where turkeys are reared—boiled nettles chopped fine are perhaps the best substitute.

At the end of a week or ten days some bread-crumbs and barley-meal may gradually be added to the egg, which may be by degrees lessened, until quite discontinued at the end of three weeks. About this time, a portion of boiled potato forms an excellent addition to the food, and by degrees some small grain may be added also—in fact, assimilating the diet very much to that of other poultry. Curds also are excellent as a portion of the dietary, but must be squeezed very dry before they are given. They are easiest prepared by adding a pinch of alum to a quart of milk slightly warmed.

By this feeding, the little chicks will get well through their *first* great danger—the tendency to diarrhœa already alluded to; and the cost of the egg will be repaid by the extra number reared.

The *second* peril to be guarded against is cold and damp: a wetting is absolutely *fatal*. The chicks should be kept entirely under the shed, on a board floor kept scrupulously clean and nicely sanded, except during *settled* sunny weather, when they may be allowed a little liberty on the grass, after the dew is quite dry. But in cold or windy weather, however fine, they must be kept in the shed, and well screened from the wind. If there be a one-storey building, their best place will be the top floor, the bottom being devoted to the sitting hens and other adult stock. Their water also must be so supplied that they *cannot* wet themselves by any possibility; and these precautions must be continued till they are nine or ten weeks old, when they will begin to "put out the red," as it is called, or to develop the singular red excrescences on the neck so characteristic of the turkey breed. This process will last some little time, and when completed the birds will be pretty fully fledged. They are now hardy, but must not be too *suddenly* exposed to rain or cold winds. Take some reasonable care of them for a while longer, and very soon they will have become the hardiest birds known in the poultry-yard, braving with impunity the fiercest storms, and even preferring, if permitted, to roost on high trees through the depth of winter. In fact, turkeys will rarely roost in a fowl-house; and a very high open shed should therefore be provided—the higher the better—the perches being placed as high as possible. They might be left to their natural inclination with perfect safety so far as their general health is concerned; but in very severe weather their feet, if roosting on exposed trees, are apt to become frost-bitten.

To attain great size, animal food and good feeding generally must be supplied from the first. By this means astonishing weights have been attained; we knew of a cock which weighed very nearly *forty pounds*, and a full-grown bird much less than thirty would stand little chance at a good show. We do not say that such weights are profitable—we believe the contrary—but we do contend that fair *good* feeding, leading to fair *good* size, is the only way to extract profit from poultry of any kind.

The ordinary domestic Turkey is of two kinds—the Norfolk (black all over) and the Cambridge. The latter is of all colours —the best, to our fancy, being a dark copper bronze; but fawn colour and pure white are often seen, as are also variegated birds, which occasionally present a very magnificent appearance. The white variety is most delicate and difficult to rear of all, but the dark Cambridge takes most prizes, and usually attains the greatest size.

We cannot here go into the question of the origin of the

VARIEGATED CAMBRIDGE TURKEYS.

domestic Turkey, or give any detailed account of the wild varieties. We can only avow our belief that a cross with the well-known American wild bird greatly improves the stamina of the young chickens, and, wherever possible, should be employed. The two races closely resemble each other, even if they are not the same; and in such cases "wild blood" is of great service. We must also allude to the surpassing beauty of the celebrated wild Honduras breed, and express a hope that it may yet be made a permanent addition to English stock. In this magnificent bird are seen in the greatest brilliancy all the colours of the rainbow, whilst in size and edible qualities it is little if at all inferior to its more sober-looking relatives. That it can be domesticated there is not the slightest doubt; and although sufficient have never yet been imported to establish the breed in Europe, we hope yet to see English yards tenanted by a bird which combines first-class merit as a table fowl, with a really Oriental splendour.

GUINEA-FOWL.—This bird, called also the *Gallina* and *Pintado*, mates in pairs, and an equal number of males and females must therefore be provided to prevent disappointment. There appear to be ten or twelve wild varieties, but only one has been domesticated in this country.

To commence breeding Guinea-fowls, it is needful to procure some eggs and set them under a common hen; for if old birds be purchased they will wander off for miles as soon as they are set at liberty, and never return; indeed, no fowl gives so much trouble from its wandering habits. If hatched in the poultry-yard, however, and regularly fed, they will remain; but must always have one meal regularly at night, or they will scarcely ever roost at home. Nothing, however, will persuade them to sleep in the fowl-house, and they usually roost in the lower branches of a tree.

The hen lays pretty freely from May or June to about August. She is a very shy bird, and if eggs are taken from her nest with her knowledge, will forsake it altogether, and seek another, which she conceals with the most sedulous care. A few should therefore always be left, and the nest never be visited when she is in sight. It is best to give the earliest eggs to a common hen, as the Guinea-fowl herself frequently sits too late to rear a brood. If "broody" in due season, however, she rarely fails to hatch nearly all. Incubation is from twenty-six to twenty-nine or thirty days.

The chicks require food almost immediately—within, at most, six hours after hatching—and should be fed and cared for in the same manner as young turkeys, though they may be allowed rather more liberty. It should be observed, however, that they require more *constant* feeding than any other chickens, a few hours' abstinence being fatal to them; and they need also rather more animal food to rear them successfully and keep them in good condition, especially in the winter. The chicks are very strong on their legs, and in fine weather may be allowed to wander with the hen when very young.

The male birds of this breed are rather quarrelsome, and very apt to beat other fowls.

The flesh of the Guinea-fowl is of exquisite flavour, much like that of the pheasant. The body about equals in size an ordinary Dorking, and is very plump and well-proportioned. Like all other finely flavoured birds, they should never be over-fed or crammed, as is sometimes done. Who would think of cramming a pheasant to make it more "fit for the table?"

PEA-FOWL.—The distinguishing characteristics of this well-known bird, are the crest or aigrette on the top of the head, and the peculiar structure of the tail covert feathers. The true tail of the peacock is short and hidden, and what we call the "tail" is, strictly speaking, an excessive development of the tail-coverts or side feathers, which occasionally have been known to extend more than a yard and a half from their insertions.

The colour of the ordinary peacock is too well known to need description. White and pied varieties are also bred, but are, in our judgment, far less ornamental. This species, called by naturalists *Pavo cristatus*, has a crest consisting of about two dozen feathers, only webbed at the very tips.

There is another variety, if possible still more beautiful, known as the Javan Pea-fowl, or *Pavo muticus*. This bird is larger than the common Pea-fowl, the male sometimes measuring more than seven feet from the bill to the end of the "tail." The naked space round the eye is also of a livid blue colour, and the feathers of the neck are laminated, or resembling scales. The most characteristic difference, however, is in the crest, which is much higher, and the feathers of which are webbed, though rather scantily, from the base, instead of being bare till near the tips. The bird also differs in only possessing his long and splendid ocellated train during the breeding season, at other times appearing with feathers not so long, and destitute of the well-known "eyes," but of a rich green with gold reflections, beautifully and regularly "barred," or "pencilled" on a very large scale, with whity-brown. This splendid bird is not very common.

A third variety has recently been described, called the "black-winged" Pea-fowl, in which the shoulders and most of the wing in the male bird are black. The hen is much lighter than the common breed, being generally of a cream colour, with a dark back. It appears a distinct race; but it must be admitted that all three varieties of Pea-fowl freely intermix with a fertile result, and so closely resemble each other in nearly all their characteristics that a common origin is not at all unlikely.

Pea-fowl are of a very wild disposition, and generally roost either on trees or on the very top ridge of a roof, to which they fly with ease. The hen lays in the greatest seclusion, and must always be allowed to select her own nest, usually deep in a shrubbery. She lays generally from five to nine eggs, but sometimes considerably more. The time of incubation is about twenty-eight to thirty days. One cock should not have more than three or four hens.

It is no use setting Pea-fowl eggs under common hens, which forsake their chickens in about two months, long before the young Pea-chicks can endure the night air. The Pea-hen goes with her brood at least six months, and the chicks *need* this. They are fed and cared for as turkeys, so far as keeping them from rain is concerned; but must be let out on the grass always in dry weather, or they will not thrive. The food is also similar in general; but some worms or other insect food should be provided in addition, in default of which some raw meat cut fine is the best substitute.

Pea-fowl are tolerably familiar, and if regularly well fed will get very tame, and tap at the window when neglected. They are, however, ill-natured, and frequently beat and even kill other fowls, sometimes even attacking children. From this cause they are ill adapted to keep in a general poultry-yard, apart from their natural impatience of restraint. Young chickens in particular the cocks will often kill, and we believe even *eat* afterwards. Their proper place is on the lawn or in the park, where the splendid hues of the cocks show to great advantage, and their peculiar shrill scream is not too near to be disagreeable.

They cannot be considered, of course, under the head of profitable poultry, being always kept for ornament. The flesh of a year-old bird is, however, excellent, and carves to great advantage on the table. Of the adult birds we have nothing to say, never having known any person who had attempted to eat one. They do not reach maturity until three years old.

CHAPTER XXIII.
PHEASANTS.

THESE birds scarcely come under the head of Poultry; but as they are often kept on account of their great beauty by amateurs, as well as extensively reared for the gun, some notice of them will not be out of place.

Confined near a house, in an aviary open to view, Pheasants will seldom lay, and scarcely ever sit. In such circumstances evergreen or other shrubs should be so arranged as to afford them some seclusion, which may induce them to breed; but it is best to hatch the eggs under a common hen. Some hen Pheasants, however, will lay and sit very well; such are usually those which have been hatched and reared in confinement, and the fact proves to our minds that with care and perseverance these birds might in time be as thoroughly *domesticated* as the other inmates of our poultry-yards. It is confirmatory of this, that whilst the wild hen only lays a dozen or fifteen eggs, in confinement, the eggs being taken daily, a home-reared bird will often lay forty or fifty, as in the case of the common fowl.

Pheasants require more than any other stock the most scrupulous cleanliness, with very abundant green food, and rather more animal substance than other poultry, otherwise the general treatment is very similar. The cock, who must be at least two years old, should be mated with three or four hens not under twelve months.

One wing should always be cut or stripped, to prevent the birds flying up and injuring themselves, as they will otherwise do. This is the more necessary, as an aviary for Pheasants should never be covered, the *adult* birds doing much better in an open run well gravelled and kept clean.

When reared as an amusement on such a limited scale, the chicks, which hatch on the twenty-fourth or twenty-fifth day, should be put out and treated generally much like chickens, or rather turkey-chicks, giving them a *board* coop made tight and sound, and only letting them run on grass when quite dry and warm; always giving them perfect *shelter* from wet and cold winds, but at the same time plenty of fresh air. They must, however, have more animal food than other chickens; and for the first few days it is best to feed entirely on hard-boiled egg chopped fine, ants' eggs, and curd pressed through a cloth till quite dry, with now and then *a little* stale bread-crumb soaked in milk. For green food, leeks or onions minced small are best. After a week their staple food may be oatmeal dough mixed very dry, and made into little pills, varied with chopped egg and bruised hemp-seed, and occasionally crushed wheat, animal food being also given. Ants' eggs, as is well known, are the very *best* animal diet for young Pheasants, and almost necessary to any great success in rearing, though much may be done without by care and attention.

The chicks must be fed for some time nearly every hour; and their water, which should always be drawn *from a spring*, must be renewed several times a day. This is the *only* way of avoiding the dreaded "gapes," which is tenfold more fatal to young Pheasants than to any other fowls; but which may be kept off by keeping the water always *clear*, and never letting them out, while young, on wet grass. Adult birds, however, are very hardy; and do not, if the soil be tolerably light and dry, require shelter from any ordinary weather, beyond what a few shrubs, or even dry brambles, thrown in their pen, will afford them.

Feeding-boxes, so commonly used, we consider bad. Keep the ground *clean*, and scatter the food broadcast. There is no better than buckwheat and barley for old birds, with green food regularly, and a little animal food now and then, like other fowls.

For rearing on a large scale, Mr. Baily, who has had great experience, recommends laying pens twelve feet square, to be erected on light dry grass land, if possible on the side of a hill facing west or south. These pens should be made of temporary hurdles or fencing, six or seven feet high, constructed of laths nailed an inch apart, and touching the ground everywhere at bottom, so as to keep out vermin. The advantages of such a plan are, first, cheapness, and secondly, convenience; as the hurdles can be taken down when the breeding season is over, and packed away in a very small compass. It is also advisable to erect them every year on fresh ground, which such a rough construction eminently facilitates.

Every such pen is adapted for a cock and three or four hens, whose wings must be cut to prevent their flying over. For a nest a slight hollow should be scooped in the ground in the centre, and filled with sand, at each end of which, and six feet apart, a short stake thirty inches high should be driven, on the tops of which is nailed a horizontal pole. Against this pole rough twig fagots are inclined from each side, forming a rough kind of shelter, which the pheasant prefers to any regular receptacle.

The eggs should be collected every evening; and if this be regularly done, every hen in the breeding-pen will usually lay at least twenty-five; the laying faculty, as we have already remarked, being increased by domestication. They are best set under Game hens, but the hen Pheasant may also be allowed a share, which she will hatch well, but is not quite so manageable with her chicks as the common hen.

The early treatment will be as already described, but when a few days—say a week—old, the board coops are placed in regular rows out on a grass-field, which should be given up to the purpose. A space round every coop should be mown close, but the rest left standing to afford the poults shelter from the heat, which they are unable to bear, suffering from it almost more than from cold. The chicks should be shut in at night, but let out strictly at *daybreak* every morning, as they are early risers.

Feeding will be as before mentioned, taking, of course, equal pains to keep the water rigidly clear. Many large breeders hang up pieces of meat to putrefy, in order to procure the peculiar white worms, called "gentles," which are collected in a tin or zinc pan placed underneath; but such should be sparingly used, as the young poults often refuse plain food after. Ants' eggs are much better.

When the breeding season is over, the old birds, and the young also when well grown, are most conveniently kept fifty or sixty together, in pens fifty feet square; being suffered to remain there until wanted, or till the breeding-pens are made up for next year.

On this system, with good management, eighty per cent. of the eggs laid may be brought to the gun, and the natural produce thus more than doubled.

Of the different varieties, the Common Pheasant is most delicate, and is rather wild. The plumage is too well known to need any description, especially as the breed is not so well adapted for the mere amateur as the beautiful Chinese or ring-necked breeds, which are daily becoming more common, and are hardier and easier to rear.

The Golden Pheasant cock is also a magnificent bird. The head bears a crest of beautiful amber-coloured feathers. The back of the head and neck is of a beautiful orange red, passing low down the breast into a deep scarlet, which is the colour of all the under parts. The neck feathers are arranged like plate-armour, and are often erected by the bird. The back is a deep gold colour, the tail covert feathers being laced with crimson: tail-feathers brown mottled with black. The hen is of a more sober tint, being of a general brown colour with dark markings.

This variety is very wild and easily startled, but is, nevertheless, more easily reared than the common pheasant, and would probably become more domesticated with perseverance in breeding under a hen. The hen pheasant herself is so shy that she scarcely ever hatches, unless in an unusually sheltered place, with shrubs and bushes arranged to resemble nature as much as possible.

The Silver Pheasant is most easily tamed of all the varieties, and is also the hardiest; whilst, in our opinion, it equals any in beauty. The cock bird of this breed has a *blue* crest, and all the upper part of the body is a silvery white, most exquisitely pencilled with fine black lines arranged with the most mathematical precision. Breast and under parts usually quite black, but sometimes a little mottled. The hen is brown, but remarkably neat and pretty.

This bird, if home-reared, may have its liberty in the poultry-yard, feeding with the other fowls; and has often been known to lay forty or fifty eggs. There appears, therefore, every reason to believe that with perseverance it might be rendered quite a domestic, and even profitable variety.

HYBRIDS between the Common Pheasant and other birds are not unfrequent. They have been known to breed with the Black Cock, Turkey, Guinea-fowl, and common domestic hen; the latter cross being not at all uncommon, as every gamekeeper knows. Such hybrids are, however, invariably sterile amongst themselves, and a very high authority* has declared them also totally unproductive when mated even with the parent; but there is undoubted evidence† of at least two birds having been reared as the produce of such a cross, mated again with the cock pheasant. The subject is only interesting from the singular fact, that although a cock pheasant is a much *smaller* bird than the domestic fowl, the cross produced is almost invariably very much *larger* in size than the mother, probably

* Mr. W. B. Tegetmeier.
† See Proceedings of the Zoological Society, 1836.

in consequence of the strong "wild blood" introduced; and hence some may think the experiment worth repeating. It is certainly true that by long perseverance great difficulties of this kind have been overcome, and hybrids, formerly considered barren, have been found at least partially fertile; but in this case interbreeding has been so often tried that we cannot consider the field very promising. One great obstacle is the extreme and apparently untamable wildness of the primary hybrid from which it is wished to breed; and the only chance of success would appear to be rearing such *singly*, in company with his or her intended mate.

We have only one further remark to make. Pheasants should never be caught with the hand, as their bones are fractured with the greatest ease. An implement should be kept for the purpose, resembling a large butterfly net, but with the bag of open netting instead of gauze. In this way they may be caught when needed with the utmost facility; but they should never be meddled with more than absolutely necessary.

CHAPTER XXIV.

WATER-FOWL.

THE above general heading, under which we shall shortly treat of Ducks, Geese, and Swans, should be borne in mind before such stock is added to the poultry-yard. They are strictly *water* birds; and although ducks may be often seen in courts and alleys where the nearest approach to a pond which they have ever known is some filthy mud-puddle, to keep animals whose proper *habitat* is so well marked in such unnatural circumstances must revolt every truly humane mind, and cannot in the long run repay any one who attempts it.

DUCKS.—In the case of these birds alone may some little

ROUEN AND AYLESBURY DUCKS.

exception be made to the above remark, as they will do well in a garden or any other tolerably wide range where they can procure plenty of slugs and worms, with a pond or cistern only a few feet across. Kept in this manner, they will not only be found profitable, but very serviceable; keeping the place almost free of those slugs which are the gardener's great plague, and doing but little damage, except to strawberries, for which they have a peculiar partiality, and which must be carefully protected from their ravages. Other fruit is too high to be in much danger.

In such circumstances there can be no doubt whatever that ducks are profitable poultry; and where numerous fowls are kept, a few should also be added, as they will keep themselves, very nearly, on what the hens refuse; but where every atom of the food they consume has to be paid for in cash, our own opinion is that ducks do not pay to rear except for *town* markets, their appetites are so everlasting and voracious. This point, however, we must leave to the experience of the reader, and proceed to consider the two principal varieties—known as the Aylesbury and Rouen. The following descriptions and accompanying remarks are from the pen of Mr. John K. Fowler, of Aylesbury, one of the largest poultry-breeders, and certainly the most successful exhibitor of ducks, in England:—

"My idea of a perfect Aylesbury drake and duck is, that in plumage they should be of the purest snow white all over. The head should be full, and the bill well set on to the skull, so that the beak should seem to be almost *in a line* from the top of the head to the tip. The bill should be long, and when viewed in front appear much like a woodcock's: it should be in prize birds of a delicate flesh colour, without spot or blemish, and with a slight fleshy excrescence where the feathers commence. If it occasionally has a very slight creamy tint it would not disqualify, but any approach to dark buff or yellow is fatal to the pen. Eye full, bright, and *quite black.*

"The legs should be strong, with the claws well webbed, and in colour of a rich dark yellow or orange. Body rather long, but broad across the shoulders, and the neck rather long and slender. The drake should have one and sometimes has two sharp curls in his tail.

"The weight of each bird in a show pen ought to be about nine pounds, but this is not very often attained.

"Immense numbers of ducks are bred around Aylesbury. It is not at all unusual to see around one small cottage 2,000 ducklings, and it has been computed that upwards of £20,000 per annum is returned to the town and neighbourhood in exchange, whilst the railway not uncommonly carries a ton weight of the birds up to the London market in a single night.

"The Aylesbury Duck often begins to lay before Christmas. Sitting hens are then procured; and immediately after hatching the ducklings are taken away from the hens and put, fifty or a hundred together, in a close *warm* place, with *one* hen tied by the leg to teach them to peck and also to huckle them. They should be given stimulating food; that is, meal well mixed with boiled meat and greaves: they are thus made fat in six or seven weeks, and if sent to market in March or April realise from 12s. to 18s. per couple.

"With regard to my own breeding stock, the selection gives me no trouble. All the large breeders know that I will give a guinea at any timefor a very fine and well-developed bird, and I thus keep my strain large, and am constantly infusing new blood.

"Many persons cannot imagine how the specimens of the breed reared *here* acquire such faultless flesh-coloured bills. The cause is local, as might be supposed. The beautiful prize tint is obtained by giving the ducks in their troughs of water a peculiar kind of white gravel found only in the neighbourhood of Aylesbury, in appearance resembling pummice-stone. In this gravel they constantly shovel their bills, and this keeps them white. Also, birds intended for exhibition are seldom allowed out in the sun, as it tans the bills sadly.

"In selecting breeding stock, drakes should be chosen with very long bills, like a woodcock's, and ducks with broad backs and large solid bodies."

For the gravel mentioned by Mr. Fowler, it is difficult to find a perfect substitute. Any other kind of clean white gravel may however be tried, and it may be well worth while for intending prize-takers to *transport* a quantity to their yards. It is also very beneficial to the paleness of the bills to let the ducks out on the wet grass in the very early morning, before the sun is up. Besides the tanning influence of the sun, it is well known that ferruginous soil has a peculiar specific effect on the bill, often turning it yellow in a single week. A bill thus stained can never be paled again; and Aylesbury Ducks should therefore never be let out on land containing iron ore.

"Rouen Ducks," Mr. Fowler states, "are reared much the same as Aylesbury, but are not nearly so forward, rarely laying till February or March. They are very handsome, and will weigh eight or nine pounds each; and, *as a rule*, do much better in most parts of England than the Aylesburys. Their flesh is excellent, and at Michaelmas is, I think, superior to the other.

"The best general description of the Rouens in plumage is to be precisely like the wild mallard, but larger. The drake should have a commanding appearance, with a rich green and purple head, and a fine long bill, formed and set on the head as I have described for the Aylesburys. The bill should look *clean*, of a yellow ground, with a very pale wash of green over it, and the 'bean' at the end of it jet black. His neck should have a sharp, clearly-marked white ring round it, not quite meeting at the back. Breast a deep rich claret-brown to well below the water-line, then passing into the under body-colour, which is a beautiful French grey, shading into white near the tail.

The back ought to be a rich greenish black quite up to the tail feathers, the curls in which are a rich dark green. Wings a greyish brown, with distinct purple and white ribbon-mark well developed. The flight-feathers must be grey and brown— any approach to white in them is a fatal disqualification, not to be compensated by any other beauty or merit. Legs a rich orange. Nothing can exceed the beauty of a drake possessing the above colours in perfection.

"The bill of the duck should not be so long as in the drake, and orange brown as a ground colour, shading off at the edges to yellow, and on the top a distinct splash or mark of a dark colour approaching black, two-thirds down from the top; it should there be rounded off, and on no account reach the sides. I may also remark that any approach to slate colour in the bills of either sex would be a fatal blemish. The head of the duck is dark brown, with two distinct light brown lines running along each side of the face, and shading away to the upper part of the neck. Breast a pale brown, delicately pencilled with dark brown; the back is exquisitely pencilled with black upon a moderately dark brown ground. The shoulder of the wing is also beautifully pencilled with black and grey; flight-feathers dark grey, any approach to white being instant disqualification; and ribbon-mark as in the drake. Belly, up to the tail, light brown, with every feather delicately pencilled to the tip. Legs orange, often, however, with a brown tinge. The duck sometimes shows an approach to a white ring round the neck, as in the drake; such a *good* judge would instantly disqualify."

To the foregoing, by far the best description of these two varieties ever published, we can add nothing. We will only remark that when intended for *fattening*, ducks should have only a trough of water instead of their usual pond, and should then be fed on barley meal. Celery will add a delicious flavour. In ordinary rearing the ducklings should be left with the hen, or mother-duck, and kept from the water entirely for a week or ten days; then only allowed to swim for half an hour at a time, till the feathers begin to grow, else they will be liable to die of cramp. They will soon be totally independent of their mother, and may then be left entirely to themselves; only taking precautions against *rats*, to which ducklings fall victims far oftener than any other poultry.

The *Muscovy*, or *Musk Duck*, appears to be a totally distinct breed, the cross between it and other ducks being, at least usually, unfertile. The drake is very large, often weighing ten pounds, and *looking* far more on account of the loose feathering; but the female is less than the Aylesbury, not exceeding about six pounds. The plumage of this variety varies greatly from all white to a deep blue-black, but usually contains both. The face is naked, and the base of the bill is greatly carunculated. The drake is very quarrelsome, and we well remember the injuries inflicted by an old villain of this breed belonging to a relative, upon a fine Dorking cock in the same yard. When excited, the bird alternately depresses and raises its head, uttering most harsh and guttural sounds, and with the red skin round the face presenting an appearance which has been justly described as "infernal."

The flesh of the Musk Duck is very good eating; but it is far inferior as a layer to either the Rouen or the Aylesbury, and cannot be considered a very useful variety.

Call Ducks are principally kept as ornamental fowl. The voice of the drake is peculiar, resembling a low whistle. They vary in colour, one variety precisely resembling the Aylesbury in plumage, but with a yellow bill, and the other the Rouen; but in both cases bearing the same relation to them as Game Bantams do to the Game Fowl. The flesh is good; but there is too little to repay breeding them for the table, and their only proper place is on the lake.

The *East Indian*, or *Buenos Ayres Black Duck*, is a most

beautiful bird. The plumage is black, with a rich green lustre, and any white, grey, or brown feathers are fatal. They should be bred for exhibition as small as possible, never exceeding five and four pounds. As they usually pair, equal numbers should be kept of both sexes. The flesh of this duck is more delicious than that of any other variety, in our estimation.

The *Cayuga*, or *Large Black Duck*, of America, is a breed well worth naturalising in this country, being hardy and a good layer. The plumage is black, approaching brown, with a white collar or neck, which with careful breeding might be soon made into a neat well-defined ring. Weight from six to eight pounds each, being thus inferior to the Aylesbury and Rouen, but with better flavour, and greater aptitude to fatten.

The Common Duck needs no description. We believe it to be the Rouen more or less degenerated, or rather, perhaps, not bred up to the perfection of that breed.

It should be remembered in keeping ducks that the *wild* birds are monogamous, and not more than two or three given to one drake, if eggs are wanted for sitting. The duck usually sits well, and always covers her eggs with loose straw when leaving them, a supply of which should therefore be left by her. The usual number laid is fifty or sixty; but ducks have laid as many as two hundred and fifty in one year; and we believe with care this faculty might be greatly developed, and their value much increased as producers of eggs. At present they are mostly kept for table.

The Aylesbury Duck is usually heaviest, and is considered the best layer by many; but on the whole the Rouen is to be preferred. At Birmingham, however, last Christmas (1866) the Rouen exceeded the Aylesbury in weight, both being shown in the greatest perfection.

Ducks should have a separate house, with a brick or stone floor, as it requires to be frequently washed down. Clean straw should be given them at least every alternate night. Other

TOULOUSE GEESE.

attention they need none, beyond the precaution of keeping them in until they have laid every morning. This is necessary, as the Duck is very careless about laying, and if left at liberty will often drop her eggs in the water whilst swimming.

Our illustration is drawn from the Birmingham prize birds of last year.

GEESE.—"Of the two principal breeds of geese," Mr. Fowler writes, "I very much prefer the Grey or Toulouse to the White or Embden, being larger and handsomer. I have had a Toulouse gander which weighed thirty-four pounds, a weight never, I am sure, attained by the White breed. They are also better shaped, as a rule, and every way the more profitable variety. The forehead should be flat, and the bill a clear orange red. The plumage is a rich brown, passing into white on the under parts and tail coverts.

"The Embden Goose is pure white in every feather, and the eye should show a peculiar blue colour in the iris in all well-bred birds."

We should recommend for market to cross the Toulouse Goose with the White, by which greater weight is gained than in either variety pure-bred; but much will depend upon circumstances. White or cross-bred geese require a pond, but the Toulouse, with a good grass run, will do well with only a trough of water, and will require no extra feeding, except for fattening or exhibition.

The only foreign varieties requiring mention are the *Chinese* and the *Canada* geese, both of which appear to be really midway between the geese proper and the swans, which they resemble in length of neck.

The Chinese Goose is of a general brown colour, passing into light grey or white on the breast, with a dark brown stripe down the back of the neck. They have much of the beauty of the swan, which they also resemble in having a dark protuberance round the base of the upper mandible. The voice is very harsh and peculiar. This breed is not a good grazer, and is best reared in the farm-yard.

The Canada Goose also is not a good grazer, and does best near *marshy* ponds, in which circumstances they will thrive and be found profitable.

With regard to the general management of geese little need be said. More than four or five should not be allowed to one gander, and such a family will require a house about eight feet square; but to secure fine stock three geese are better to one male. Each nest must be about two feet six inches square, and, as the goose will always lay where she has deposited her *first* egg, there must be a nest for each bird. If they each lay in a separate nest the eggs may be left; otherwise, they should be removed daily.

Geese should be set in March or early April, as it is very difficult to rear the young in hot weather. The time is thirty to thirty-four days. The goose sits very steadily, but should be induced to come off daily and take a bath. Besides this she should have in reach a good supply of food and water, or hunger will compel her, one by one, to eat all her eggs. The gander is usually kept away; but this is not very needful, as he not only has no enmity to the eggs or goslings, but takes very great interest in the hatching, often sitting by his mate for hours.

The goslings should be allowed to hatch out entirely by themselves. When put out, they should have a fresh turf daily for a few days, and be fed on boiled oatmeal and rice, with water *from a pond*, in a very shallow dish, as they should not be allowed to swim for a fortnight, for which time the goose is better kept under a very large crate. After two weeks they will be able to shift for themselves, only requiring to be protected from very heavy rain till fledged, and to have one or two feeds of grain daily, in addition to what they pick up.

For fattening they should be penned up half-a-dozen

together in a dark shed and fed on barley meal, being let out several hours for a *last bath* before being killed, in order to clean their feathers.

"For exhibition," Mr. Fowler says, "all geese should bo shut up in the dark, and fed liberally upon whole barley or oats thrown into water. It is essential to great weight to keep them very quiet, letting them out in the water, however, for half an hour every day."

SWANS.—There are six or seven varieties of swans known to naturalists, but only three are at present, or likely to be, domesticated in this country—viz., the large English White or Mute Swan; the Australian or Black Swan, and the Chili or Peruvian Swan. The plumage of the two first needs no description; but that of the Chilian Swan differs from either in being white on the body, with a black head and neck, making rather a pleasing contrast of colour. In size the White Swan is largest of all. All three varieties are long-lived, and particular birds are reported to have reached the age of one hundred years.

The following remarks on swans are by Trevor Dickens, Esq., of London, who is well acquainted with these beautiful birds :—

"Besides ornament, swans are often of considerable use in clearing lakes or canals from weeds generally, and in particular from the one peculiar plant which within late years has become an only too well-known nuisance. To this there is, however, a drawback, as they also destroy the young fry of fish.

"The large English White Swan is most beautiful in form, as well as in colour. The Black Swan is also apt to be bad-tempered, and is more mischievous on the water; for all which reasons the first place must still be given to the magnificent old English breed. It sometimes occurs wild, but in such circumstances is always of a rather grey colour instead of pure white. The finest swans in England are to be seen in the

Thames and Trent rivers, and at Abbotsbury in Dorsetshire. The Marquis of Exeter, at Burghley Park, the Marquis of Abercorn, and in Scotland the Earl of Wemyss, are also well known for their beautiful swans.

"The female swan lays in February, every other day until seven to nine eggs are laid, and then sits for forty-two days.* More than five cygnets, however, are seldom hatched. The nest is made somewhere amongst the flags and weeds at the water's edge, and it is dangerous to approach either the male or female during incubation, as they are very irascible, and a blow from their strong pinions will even break a man's arm.

"The cygnets are best fed by throwing meal upon the water. The old birds, if they have a large water range, will only need feeding in severe winter, when they should have grain. They also like grass thrown to them, and bread, which they will frequently eat from the hand.

"It is usually asserted that the swan is strictly monogamous. But I have frequently seen *two* females with the male during the breeding season, and believe the idea to have arisen from the stronger female always seeking to drive the weaker away before breeding. Full-grown males never agree at all, and must, therefore, be kept separate."

It is impossible to add anything on the general management of swans, as the young birds must be left to shift for themselves, the parents being too jealous and powerful to submit to restraint. But for this, they might perhaps be more widely kept, as the young cygnets are excellent for the table, and very easily reared.

* Bechstein, a most accurate observer, and many others, contradict this, and state that the swan sits for only thirty-five days.

SECTION V.

THE HATCHING AND REARING OF CHICKENS ARTIFICIALLY.

THE HATCHING AND REARING OF CHICKENS ARTIFICIALLY.

CHAPTER XXV.

THE INCUBATOR AND ITS MANAGEMENT.

THE artificial hatching of chickens, as is well known, has been practised as quite an ordinary thing in Egypt for thousands of years, and with the most complete success; yet, strange to say, is only a very modern experiment in Europe, and, on the whole, by no means a satisfactory one..

To give a history of all, or even of the principal attempts that have been made to hatch chickens by heat artificially applied, would far exceed our limits, and would he of no practical use. It will be enough to say that Reaumur was the first who really took the matter up in earnest, and he succeeded also about as well as those who have come after him. His method was to place the eggs in wooden casks, or other vessels, and then to surround the whole with fresh dung in a state of fermentation, which was renewed as often as necessary. For obvious reasons this system is never likely to be popular; but it is mentioned by Mr. Geyelin as still employed with success in France.

Cantelo was the first to imitate the hen in supplying the heat from *above*, and his apparatus was very fairly successful, the only real objection being its great cost. Precisely the same may be said of the elaborate contrivance of Minasi; it hatches chickens with success, but is too costly ever to become popular, unless the price can be greatly reduced; and, on the whole, the only incubators we consider well adapted to general use are those of M. Carbonnier, Mr Brindley, and Mr. F. Schröder, which we shall first describe, referring afterwards to the essentials of successful management. That described by Mr. Geyelin in his well-known pamphlet we do not think worth consideration.

M. Carbonnier's incubator is so simple as to be easily understood without a diagram, and can be constructed by any country workman. The heating apparatus consists of a tin or copper cistern, or boiler, of any desired size, made with a flat bottom, and heated by a lamp, for which a chamber is provided in one end. The lamp must, of course, be constructed to burn for a certain time without alteration, and it is essential that the lamp chamber be in the *end* of the cistern, that there may be a proper and regular circulation of the water. The cistern should be kept nearly filled, with a thermometer constantly immersed to show the temperature.

Under the cistern slides a drawer, in which the eggs are placed upon a little hay. They should not, however, be exposed direct to the heat of the cistern—the great failing of most incubators—but ought to be covered with a piece of canvas, on which is spread a layer of sawdust half-an-inch thick. The sawdust readily becomes warmed by the heat of the cistern, and, resting gently upon the eggs, warms them in a more natural manner than any other incubator we know. In the egg-drawer a second thermometer should be kept, to show the heat to which the eggs are actually subjected. It should be observed that in this, as in every other incubator, the cistern must extend some inches beyond the eggs on every side, or those outside will not get their proper heat, and therefore perish.

The management of this incubator is very simple. The

THE INCUBATOR.

Fig. 14. Brindley's Incubator.

A A. Temporary Artificial Mother for newly-hatched Chicks.
B B. Lamp and Reservoir.
C. Egg Drawer.
F. Hot-water Boiler.

lamp must be so adjusted that the actual temperature of the sawdust may be kept at a standard of 102° or 103°,* and then

* As this temperature varies from that usually given, see remarks on the subject further on.

regularly and properly attended to, so as to ensure this. Once a day the eggs must be withdrawn, and exposed for twenty minutes to the cold air of the apartment; and, when replaced, each egg must be turned over, and the sawdust laid again upon them, and sprinkled, from a small watering-pot, with water heated to 105°, so as to make it slightly moist. In all these proceedings Nature is most exactly followed, and the result will be a good proportion of well-hatched chickens.

The arrangement of Mr. Brindley's incubator is shown by Fig. 14. F is a copper boiler, heated either by a gas jet or by a paraffin lamp, B, furnished with a reservoir, also marked B, carefully constructed to burn with steadiness. From this boiler the hot water flows constantly through a system of metal pipes arranged in a horizontal plane between two plates of glass, which thus form a *hot-air chamber* heated by the pipes. Under the lower glass plate slides the drawer, C, lined with felt, which contains the eggs, E. At each side of the lamp, at A, are temporary receptacles, or artificial mothers, to receive the chickens for the first day, after which they must be removed and provided for separately. The hot-air chamber is provided with a "safety valve," acted on by the expansion of mercury, which can be balanced to open at any desired temperature. Such a valve appears to have been first employed by M. Vallée, of the Jardin des Plantes, Paris; but we believe Mr. Brindley's valve to be superior, and, within *reasonable limits*, to answer its purpose very fairly. To make any valve the *sole regulator*, and expect it alone to keep the heat uniform, as some appear to do, is absolute nonsense. All that can be expected of any valve is to open when the heat becomes two or three degrees too high, and admit cool air to reduce it to the proper temperature; but if the air be carelessly allowed to get really *hot*, the valve, though open, cannot keep the heat down, neither can it guard against a *lower* temperature than is proper.

Mr. Brindley's incubator, it will be seen, differs radically

THE INCUBATOR. 207

in principle from the preceding, as also from the next we shall notice, in that water is not employed *directly* to warm the eggs, but simply to impart heat to a chamber of hot *air*, through which the heat is communicated. Otherwise the management is very similar. The eggs require to be withdrawn and cooled once a day; and before they are replaced they should be carefully turned, and sprinkled with warm water, which should

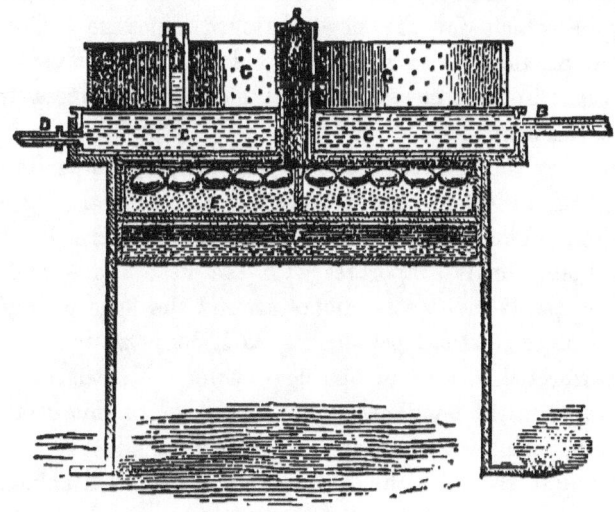

Fig. 15.

also be allowed to moisten the felt lining of the tray in which they are contained.

The last incubator we shall describe, and the last yet made public, is the invention of Mr. F. H. Schröder, the able manager of the National Poultry Company, and is shown in section in Fig. 15. Mr. Schröder has adopted an altogether distinct and separate boiler, which is not shown, and which is connected with the hot-water tank, C, of the incubator by two pipes, B being the inlet pipe and D the outlet. This tank is provided with an open tube, I—in which a thermometer can be placed to show the temperature—and with a ventilating tube, H, which is open at top and bottom. Under the tank slide the egg-drawers, E, which in area resemble the quadrant of a circle, Mr. Schröder's incubator being of a circular form. The bottoms of these drawers are of perforated zinc. Under all is a tank, F, of *cold* water. The space, G, above the hot water tank, is surrounded by perforated zinc, and partly filled with sand, both to preserve the heat, and to form a convenient and warm receptacle for the newly-hatched chickens. Curtains are also provided to surround the sides of the incubator, and thus guard in some measure against change of temperature in the apartment.

In using this incubator the egg-drawers, E, are partly filled with chaff, or other similar material, on which the eggs are deposited. The water from the cold water cistern F, underneath them, slowly evaporates with the heat above, and preserves a gentle moist atmosphere around the eggs during the process of incubation, percolating as it does through the chaff and perforated bottom of the egg-drawer. Ventilation takes place through the middle shaft, or pipe, H. In this incubator, therefore, sprinkling the eggs is not needful, all that is necessary being to replenish the cold water tank, F, when exhausted; but the eggs, as in *all* incubators, should be withdrawn, cooled for half an hour, and afterwards turned, every day.

We have no hesitation in pronouncing the cold water tank in this incubator a most valuable invention, and one which answers its immediate purpose well; while it also, to some extent, tends to equalise the temperature. The arrangement at top for the chickens is also very simple and convenient, and the whole shows both originality and ingenuity in a very high degree.

That artificial incubation will ever *commercially* supersede, in ordinary seasons and for ordinary eggs, the natural process, we do not for a moment believe. That it does so in Egypt is not the slightest argument; in that country there is a climate both warm and steady, whilst in this it is both cold and very variable. The value of incubators is to hatch when hens *cannot be had*, and in such seasons 70, 60, or even 40 per cent. will often be thankfully accepted by breeders for exhibition as ample return.

Now it will not do to purchase an incubator, light the lamp, put in the eggs, and expect that, provided the lamp be only kept burning, all will go right. The consequence would be utter failure. And, on the other hand, we would undertake to hatch somewhere between the averages we have quoted with the very *worst* Incubator that was ever constructed; only perhaps changing the lamp, if very faulty, for one constructed to burn more regularly. Certain precautions must be taken, certain conditions must be secured, and certain errors must be guarded against.

And first it must be remembered that in artificial hatching it is *absolutely* necessary the eggs be fresh. Hens will hatch eggs a fortnight old or more—incubators scarcely ever. Of course, if the artificial process were perfect, this difference would not be. But it is *not* perfect—it is a substitute. We are fighting against a host of difficulties; we must, therefore, take the fact as we find it, and choose only eggs that do not exceed five or six days old. This caution *cannot be neglected with impunity;* if any inventor promise otherwise, let the credulous purchaser only try.

Again, the incubator must be placed where it shall not be exposed to jarring or concussion. That timid hens always hatch small broods is well known; yet many appear to think that they can expose their *artificial* hen to any vibration or noise without injury. This is to court a danger which Nature is ever seeking to avoid.

That the eggs should be daily *sprinkled* has already been mentioned. Only in Schröder's Incubator can this precaution

o

be dispensed with; and we cannot but consider that gentleman's evaporating tray the most valuable feature in the whole invention. Still it answers quite as well to sprinkle with water daily, *if it be done;* the value of Schröder's plan is in the case of forgetful operators. It must be remembered that eggs in the circumstances we are considering require moisture more than under a hen in the very driest season, since even then eggs naturally hatched get a *little* humidity from the perspiration of the hen's body. But in an incubator *all* must be supplied, and any omission is death and failure.

But the greatest mistake is in seeking *too high a temperature*. In every published work we have seen, the standard and proper heat for the eggs is given as 105°, and we have not the slightest hesitation in saying that to this the largest proportion of failures is due, the chickens being roasted in their shells. We do not mean to say that 105° will kill the chicks, or will not hatch them; but we do say that some hours of 108° will kill *a few;* and as in this climate it is *impossible* to maintain a constant temperature, if 105° be taken as the *standard*, it is sure to be exceeded again and again; and thus, two or three perhaps at a time, the chicks are killed. On the other hand, it has been conclusively proved that whilst 98° is not enough to hatch successfully, the temperature may be allowed to sink so low for some time occasionally with little injury. Let 102° therefore be taken as the proper standard for the eggs, and more chickens will be hatched than have ever been. A rise of several degrees will then not be fatal, whilst an occasional fall will also be borne; and, with fresh eggs, a good hatch may be expected.* And this leads us to the great difficulty of all artifi-

* Since writing the above we have had a communication from Mr. Brindley, in reply to a note embodying the above opinion, in which he fully concurs with the view we have here expressed, and encloses the directions issued with his patent incubator, in which 103° is given as the proper temperature. We are happy to find our judgment thus corroborated, and willingly give him credit as the first to publish a correct statement on the subject.

cial hatching—that of maintaining a *regular* temperature in our variable climate. The same lamp-flame will not keep up during the night the same heat in the water by many degrees as it maintained during the day, and the difference must be carefully provided for, or disappointment will ensue. This is where many fail, and where so much attention is requisite. Changes of weather must be guarded against, and compensated in like manner; and for all this there must be the most constant reference to the thermometers, both the one in the heating chamber or cistern, and the other which should always be kept in the egg-drawer itself. It is here that Mr. Brindley's valve will be useful; but it will not do to *depend* upon it; it will *help*, but it will not do the work of supervision. Mr. Schröder's idea of surrounding the whole with curtains is also good, and may be applied to any incubator. But, with all these helps, the lamp itself must be carefully arranged so as to give more heat during the night than in the day, and in cold weather than in mild; and the process should also be carried on in the part of the house where the temperature is most uniform. A bedroom is a good place, as it is untenanted in the day, whilst at night the occupants help to keep up the heat. Another, and the best plan, is to place the incubator in a room with a fireplace, but not near it, and to light a fire in the evening proportionate to the coldness of the weather. By this means something like uniformity may be preserved in *the room*, and this will go a long way to maintain it in the machine.

It is for the same reason that in the simpler forms of incubators the hot water cistern should extend several inches beyond the eggs on every side. In *small* machines this is specially required; and the neglect of so necessary a precaution is one great reason why the small ordinary incubators frequently purchased almost always fail; the outside eggs cannot be kept warm enough without roasting the others.

It is by constant and careful attention to such minute

circumstances, and *thus only*, that success in hatching can be attained. No particular form of incubator will answer *without* such care, and *with* it almost any will do, though the three we have selected are indisputably the best. The two last men-

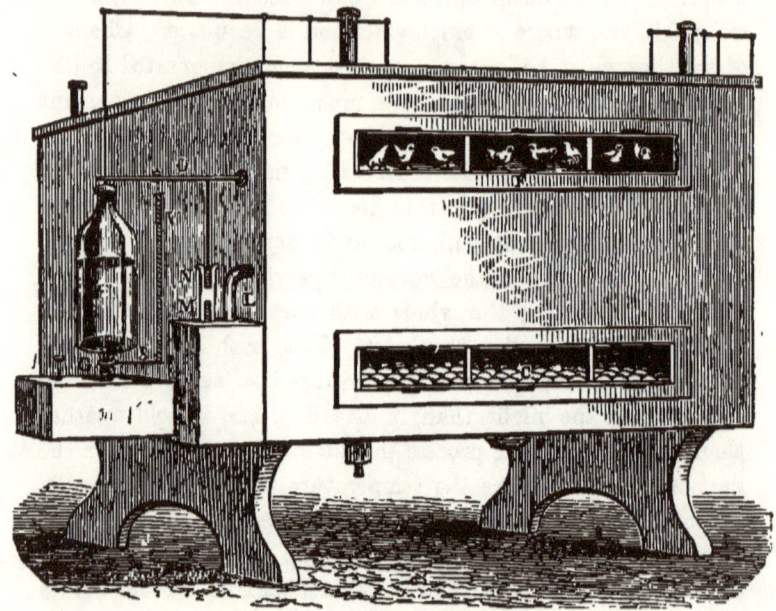

tioned are more elaborate, and perhaps more complete; whilst that of M. Carbonnier is the cheapest and most natural. In his system we particularly like the layer of damp sawdust gently *resting* upon the eggs, and communicating a moist heat from the hot cistern, which closely approaches the natural hatching of a hen, and we believe will be ultimately found to be more successful at the very last, when the chick actually *chip the shell*, than any other.

Since the preceding pages were first published, the incubators there described have been made and sold to some extent, but success has been but limited with them all, owing chiefly to the fact that very few people have patience to give that constant care which *alone* can purchase success. Lately an incubator has been extensively sold by Messrs. Jacob Graves and Co., of Boston, U.S., which far surpasses all those here described in the perfection of its apparatus for preserving a uniform temperature. Its external appearance is represented in the engraving opposite, and it has had a greater amount of *actual success* in work than any other ever yet made. At the Boston show, in 1873, it hatched regularly and successfully, challenging the admiration of many fanciers who had previously lost all faith in "artificial hens."

We have reason to believe that an English incubator is, as we write, nearly perfected, which will remove most of the *practical* difficulties hitherto found so serious, and enable eggs to be hatched without more than a moderate amount of trouble. This measure of success has been attained by attending to conditions we long ago pointed out as 'essential, and especially by affording to the eggs a constant and ample supply of *fresh air*, in which most incubators hitherto constructed—even the best—have been deficient.

CHAPTER XXVI.
REARING CHICKENS ARTIFICIALLY.

THE artificial rearing of chickens must be regarded as a question entirely distinct from the artificial hatching of them, and may often become advisable, or even necessary, when they have been hatched under a hen. The mother may die just when her care becomes most necessary; or she may be a valuable hen, whose eggs are much wanted, and whom it is not advisable to subject to the wear and tear of a young brood. And lastly, some persons consider that it is absolutely *better* to bring up chickens by hand, even when they have been naturally hatched; believing that under the shelter provided, and not being forced to accompany the hen in her rambles, a greater portion are reared, that they grow faster, and make ultimately finer fowls.

We cannot certainly agree in such an opinion, though there are respectable authorities who hold it. We admit that, with care, chickens may be reared with *as much* success as by a hen, but more we cannot concede; and even for this much the greatest care is requisite, and proper management is absolutely necessary.

Some sort of an "artificial mother" must of course be provided, and the best form of all is the ordinary one. This consists of a board sloping down from four inches above the ground to about two inches; and for a brood of a dozen chicks, about a foot square. It is covered on the under side with a piece of lamb or sheep-skin dressed with the long wool on, and which should only be tacked round the *edges* of the board, so as to fall a little slack with its own weight, and thus rest upon the chickens. By attending to this, as well as to the slope of the board, the largest and smallest chickens will be accommodated with equal comfort. A few small gimlet holes should be bored in this cover for ventilation.

Instead of sheepskin, some employ a manufactured article which resembles a number of cotton wicks hanging thickly from a sort of linen foundation. We should *prefer* this when obtainable, but it is very difficult to procure, while sheepskin is always at command.

The board so furnished must be mounted on two sides and a back of wood, the back being two inches high, and the sides, of course, sloping up from that height to four inches in the front, which is left open for the chicks to enter by. This front side is, however, furnished with a curtain of flannel four and a half inches deep, which thus sweeps the ground and excludes the cold air, whilst the chicks push under it either way with the greatest ease. There should be *no bottom* at all. We believe the addition of a wooden bottom to be the great reason why so many have difficulty in rearing chickens artificially. Such a bottom may be sanded or covered with ashes with the most sedulous care; but it *will* harbour vermin, and become more or less tainted, and the chickens will then be sure to droop away. Moreover, it is hopeless to expect good constitution in birds reared more than the first fortnight on a wooden floor. Let the "mother" be set on the *ground*, evenly covered an inch deep with sand or nice dry ashes; let it be never left two nights in precisely the same spot, and let the ground it is to occupy be

Fig. 16.

A is the frame of the wire run.
B are the wire blinds, each movable, and thus allowing the run to be cleaned out easily.
C is the "hood," which takes off—as shown at the dotted line—and is used when the chickens are able to perch.
D is the hairy cover—the substitute for the hen's body. This is detached, and fits either along the dotted line, and so is suited for chickens not able to perch, or when the hood and perch are used, forms the cover to the same.
        E Perch.   F Tressels and stand.

perfectly clean and dry before each removal. Such care will be well repaid.

In severe weather, however, it is almost necessary to keep the chickens within doors till about three weeks old, and a wooden floor to the "mother" then becomes necessary. No better arrangement perhaps can be devised than that contrived by Mr. F. H. Schröder, and shown in Fig. 16; which is constructed to stand upon trestles at a convenient height for cleansing. The roof of the "mother" is here made so as to be raised at pleasure when the chickens are able to roost, and allows of a perch being introduced; but long before this time they ought to be removed to the ground, if designed for anything but mere in-door amusement. The floor of such temporary homes must be scrupulously cleaned every day, and sprinkled with clean sand or fine ashes so as completely to cover the wooden bottom.

But in ordinary weather it is better, warming the "mother" with hot water, to put the chickens on the ground at once. In front of it must be a covered run, which may be about three to four feet long, enclosed at the sides and end by board, and covered with glass. The board enclosing it must not be *less* than a foot in height, with a few holes bored near the top for ventilation; otherwise the atmosphere within will be too close for the chicks to live in it. It is well to make the glass top so that it can be lifted in warm weather like a cucumber frame, or the heat will become stifling. Neglect of these precautions also causes many failures.

In front of the covered run, again, must be an open run fenced in and covered over with small mesh wire netting. This may be any convenient size, and should extend over grass if possible. Communication between this open run and the covered run and "mother," is maintained by one or two small traps large enough for the chicks to pass when tolerably well grown, which are left open to allow of their free passage in fine weather, but should be kept closed when it is wet or very cold.

As in very heavy weather the glass roof of the covered run is not sufficient protection, the whole arrangement must be placed under an open shed in some sheltered situation.

MANAGEMENT OF THE CHICKS.

Cleanliness in the two runs is of nearly as much importance as under the "mother." They should be raked over constantly, if gravel or sand; and if set upon grass, the whole should be moved to fresh ground every two or three days.

The fleece or upper part of the "mother" itself is liable, if neglected, to get infected with insects. To prevent this, powdered sulphur should be frequently dusted into it, and a little paraffin put on here and there occasionally will also in a great measure expel them by the strong smell. No point is perhaps so universally neglected as this. But chickens when tormented by vermin *never thrive*, and we believe are occasionally worried even to death by this intolerable plague.

Such will be all the accommodation needed in ordinary summer or spring weather, during which the chicks, when in the "mother," will keep *themselves* comfortably warm. But for the preservation of broods hatched in January or February, it will be necessary to add artificial heat, which may be done by having on the top board of the "mother" a vessel to be filled with hot water the last thing at night, and once or twice during the day. In very severe weather even this will not be sufficient, and the water must be kept hot through the night by a lamp or other contrivance. Of course, if there be hot air apparatus for a greenhouse, or any other permanent source of heat, it may be made available in any convenient manner, and a lamp dispensed with.

The feeding will not differ from that already given. Hard boiled eggs chopped up, and *very coarse* oatmeal moistened with milk or water, is best to commence with, as the chickens will begin to peck much more readily at such tiny *morsels* than at anything in the shape of sop. Groats chopped up small are also very useful in teaching them to feed. This is, in fact, the only difficulty, and is best got over by tapping on the floor with the end of the finger, at the same time clucking like a hen. But very few chickens give any trouble in this way, and the art of feeding is one which, once learnt, is fortunately never forgotten. Let not animal or green food be neglected, or the chickens will never be superior specimens; and let grain be added by *degrees*, but still letting the chief diet till at least three months old consist of soft food. This, however, has been fully treated of in Section I., and we will only add a caution that the young birds be never *neglected*. Remember that chicks with a hen, if at liberty, can almost always procure *some* food— enough to maintain life at least—if their regular meal be forgotten; whilst those reared in this manner are *entirely* dependent upon their owner's care, and one forgotten meal, even if not fatal at the time, frequently lays the foundation of mortal disease, by leaving the poor little things with no strength to endure any inclemency of the weather. The want of such support is what makes bread sops so objectionable a food for young birds.

To sum up all : WARMTH (with ventilation), CLEANLINESS, and CONSTANT FEEDING will give unfailing success in the rearing of chickens artificially; and when there has been signal failure, tho cause will be found in neglect of one of these three. The whole art is therefore simple enough, and every large poultry-keeper should make himself to some extent conversant with it, as such experience may often prove serviceable, even should he be one of those who shun "incubators" as they would the plague. For instance, a hen cannot cover well more than six or seven chickens if hatched very early, but can *hatch* well ten or eleven: hence a poultry-breeder-experienced in artificial rearing has much advantage over another ignorant of it, as he can set all his hens in January (when "broody" hens are very scarce) on their full complement of eggs, and when hatched give each as many as she can properly protect, and bring the remainder up by hand. To exhibitors especially the possibility of thus getting early stock in increased numbers is of great and special importance.

SECTION VI.

THE BREEDING AND MANAGEMENT OF POULTRY UPON A LARGE SCALE.

POULTRY ON THE LARGE SCALE.

CHAPTER XXVII.

SEPARATE ESTABLISHMENTS FOR REARING POULTRY. POULTRY ON THE FARM. CONCLUSION.

IN seeking to give such information as may be useful to any contemplating the wholesale rearing and keeping of fowls as a distinct business, we labour under the great disadvantage that there is no successful concern of the kind in England to which we can refer. That this is not for want of a market for either eggs or chickens, is proved by the continuous high prices of the one, and the many millions of the other yearly imported from France and Ireland. Still it is the fact;* and for any actual examples which we can consider worthy of imitation, we have therefore been compelled to cross over to France, where such enterprise is carried on to an extent, and with a success little dreamed of in this country, and which proves that here also the first who shall bring to bear upon it the same amount

* In this and other observations which more or less directly appear to reflect upon the well-known National Poultry Company's establishment at Bromley, it is not meant to assert that the concern there is a losing one; on this point we have no information whatever, and make no such imputation of the slightest kind. But it has become, from sheer necessity, a mere assemblage of pens for breeding and showing prize poultry, and selling eggs therefrom; and has altogether failed to provide a supply of fowls for *the market* at a cheap rate, as every one predicted it would, and on which ground it was ostensibly inaugurated.

of practical knowledge, sound judgment, and good business management, will not fail to reap a similar harvest.

It is, however, very necessary to make these reservations. Nothing is more easy than to publish sanguine calculations showing from one to three hundred per cent. profit to be derived from such concerns, and more than one such have we seen; but unless these computations are founded upon some sound practical knowledge of such details as are contained in the foregoing pages, they cannot but prove delusive. It was here that Mr. Geyelin so signally failed. With many good ideas—some of which have been found truly valuable—he utterly lacked that real *knowledge of fowls* which could alone have turned them to account; and hence his well-known pamphlet, full as it was of really useful conceptions, and awakening as it deservedly did very great attention to the subject of wholesale poultry-breeding, abounds also with absurdities which could only provoke a smile from every one who had actually kept fowls. He was essentially a theorist; and since his theories involved certain principles which were fundamentally wrong, that his plans should fail practically was an inevitable necessity. And that they have done so is an admitted fact.

Mr. Geyelin's fundamental idea was, that with proper care and judicious feeding, fowls could be bred, reared, and kept for any purpose—either for chickens or for eggs—far more economically, and in better health, in close confinement, than even with a moderate degree of liberty. And to those ignorant of the subject he apparently demonstrated his point. He alleged truly that the chickens would be protected from wet and cold; that they would never be over-tired; and that they would always be properly fed; and in his arrangements he therefore provided that they should be hatched and reared on wooden floors. But he *forgot* that such treatment would not give *constitution*, without which no system can in the long run be remunerative; and this one flaw in the argument has rendered valueless all his after reasoning. On the first appearance of Mr. Geyelin's pamphlet, we ventured to predict that whilst he might keep in health and good condition grown fowls, his plans would fail altogether with regard to chickens; and since chicken-rearing is at the very root of all plans for keeping poultry on the large scale, would practically fail altogether. The event has justified this prognostication to the letter; for whilst the National Poultry Company have kept in good health, and taken numerous prizes with, adult birds from their small pens, they have not succeeded in sending any amount of dead stock to the London market ; and on a recent visit to their establishment at Bromley, we found the Geyelin system of rearing the broods altogether abandoned, and the chickens were being brought up out of doors as usual. At a smaller establishment in the provinces, built on the exact model of that at Bromley, we found precisely similar results.

Such being the case, we shall not give any detailed description of Mr. Geyelin's plans, referring those who may be desirous of investigating them to his own pamphlet for further information.* But in justice to him we must nevertheless remark that he has rendered real assistance to the advancement of poultry-breeding of no small value. He has conclusively proved that adult fowls can be kept in health in pens of only six feet by twelve, and demonstrated in connection with this the great value and importance of deodorisation ; he was the first in this country to insist *publicly* upon the necessity of giving soft food as well as grain ; and, most important perhaps of all, he pointed out perspicuously the design of nature, and the necessity to the *most* profitable result, of making the fowls feed the land whilst the land fed the fowls. These are im-

* Since these pages were written, the disastrous failure of the National Poultry Company has more than confirmed all our remarks.

portant services, and it would ill become us not to acknowledge them, though we cannot follow him to his conclusions.

In attempting ourselves to give such information as may be useful to those contemplating this branch of commercial enterprise, we shall in the first place, translating from an interesting work* published under the authority of the French Minister of Agriculture, give a short and illustrated description of one of the latest and best managed establishments in France, afterwards making such remarks as may appear advisable.

The establishment in question is said to belong to the Baroness de Linas, and is situated at Charny, a village near Paris. Left a widow some years since, with a small estate of about fifteen acres, which bears the name of Belair, Madame de Linas, partly for amusement and partly in order to augment a rather scanty income, turns her attention to poultry, and has for some time succeeded in both objects. Many of her arrangements are peculiar; but all are the result of much thought, and are worthy of attentive examination.

The poultry-house at Belair is represented in perspective by the accompanying plate. It is in two storeys, each $7\frac{1}{2}$ feet in height; measures in all 60 by 15 feet, and is divided by partitions into four compartments of equal size. This house is designed for the accommodation of about 1,200 laying hens, with a due proportion of cocks, which are lodged in the four upper apartments; whilst the lower are devoted to storing, cooking, hatching, and other necessities of the business.

Round the front and ends of the house there is a gallery, five feet wide, at the level of the upper floor, roofed like a verandah, on which the doors of the fowl-houses open, and to which the birds ascend by broad step-ladders. The gallery carries a small railway, travelled by a truck, and at each end is a lifting tackle, by which simple means the manure and eggs are col-

* "Poules et Œufs," par Eug. Gayot. Paris: Librairie Agricole, 26, Rue Jacob.

THE POULTRY-HOUSE AT BELAIR.

lected and lowered down, whilst straw, sand, and anything else required, are hauled up, and distributed with the least possible amount of labour. The doors of the hen-houses do not open on hinges, but slide in panels, so as always to leave the gallery clear. They are furnished with traps, as usual, by which the

Fig. 17. View of Hen-house.

birds can enter when they are closed. The object of thus elevating the hen-houses are two-fold—dryness and salubrity, and security from thieves and vermin, as the ladders can be taken away at night, and all access cut off.

The interior arrangements of all four upper apartments, or hen-houses, are precisely similar, and are shown in Figs. 17 and 18.

Each apartment is designed for about 330 fowls, and the interior dimensions are 16 feet by 15. The perches, shown in plan at J, Fig. 18, consist of flat planks, four or five inches wide, with only the top corners rounded off, and arranged on a frame so as to be movable, at a height of 16 inches above the

Fig. 18. Plan of Hen-house.

floor. Such perches never cause crooked breast-bones; the heaviest fowl can reach them, and there is never any dispute for the highest place, which is always the case when arranged *en échelon* or ladderwise.

The nests, shown at N N, are arranged in five tiers against the front and back walls. They are formed very simply, by dividing long square troughs, open at the top, into compartments, by means of partitions sliding in grooves. The bottoms of the troughs project, so as to form broad ledges, along which the hens can walk; and inclined ladders, shown at E, give ready access to each ledge, and, consequently, to any nest.

The floor is formed of resinous pine wood, in order to repel vermin. Every crevice is stopped up, and the whole scraped clean and profusely sanded every morning whilst the birds are at their first meal. In addition to this, the whole is well fumigated and whitewashed twice a year.

Air is admitted to each apartment by the pipe B, which rises through the middle of the floor, and which is brought from over the furnace in the kitchen, as shown at O in the plan of the ground-floor (Fig. 20). By this means the temperature in winter is kept warm. Another pipe through the ceiling carries off the products of respiration. In summer, ventilation is further promoted by keeping open the Venetian blinds, F, with which the house is furnished. A is the door opening upon the gallery.

Fig. 19 shows the arrangement of the open runs, which occupy about an acre and a-half each, and are of a wedge-form, converging on the compartments of the fowl-house, and opening towards the further end, where they are bounded by a clear running brook. Each run is provided with a spacious shed, built on rising ground, and small clumps of trees and bushes are also grown, to afford shelter from the sun. Shallow pits, filled with fine sand, are also provided. Every three months a fourth of each run is sown with hay-seed, and lightly dug over, in order to renew the turf and bury all manure. A supply of worms is also in this way afforded to the fowls.

The fences might, of course, be of any adequate kind, but are constructed at Belair in a very ingenious manner. A double row of poplar, elm, or apple-trees is planted, and suffered to grow for several years unmolested. Then each tree is nearly cut through with a bill-hook, and bent over, but

Fig. 19. THE OPEN RUNS FOR CHICKENS AND FOWLS.

230 POULTRY ON THE LARGE SCALE.

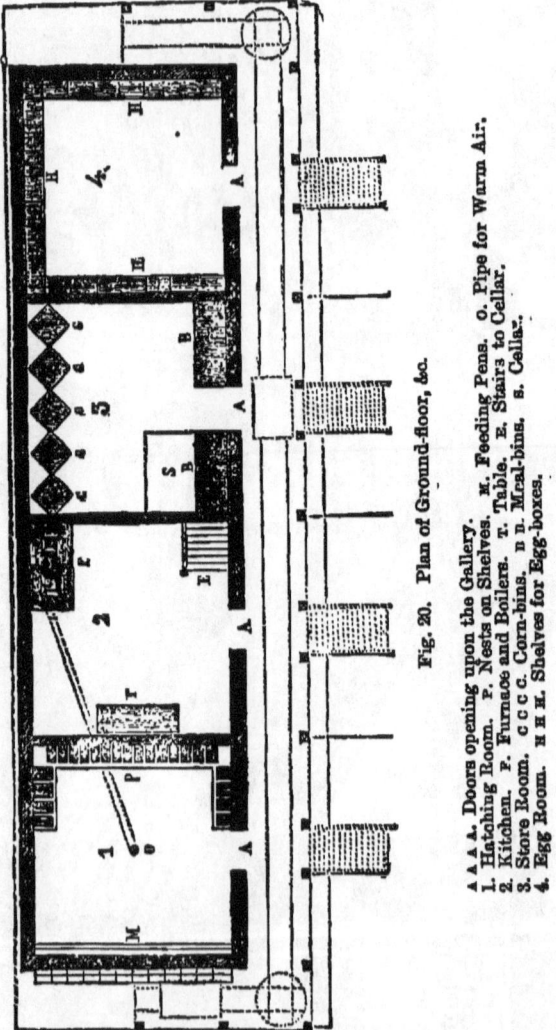

Fig. 20. Plan of Ground-floor, &c.

A A A. Doors opening upon the Gallery. M. Feeding Pens. O. Pipe for Warm Air. 1. Hatching Room. P. Nests on Shelves. R. Table. E. Stairs to Cellar. 2. Kitchen. F. Furnace and Boilers. T. Table. E. Stairs to Cellar. 3. Store Room. C C C C. Corn-bins. D D. Meal-bins. S. Cellar. 4. Egg Room. H H H. Shelves for Egg-boxes.

leaving, of course, some of the wood and a broad strip of the bark. The effect of this treatment is to make the trees send out vigorous shoots in every direction, of which the largest are again cut and laid down as before; and the whole being kept in bounds by a rude trellis, the effect in a few years is a dense living wall of foliage, which is absolutely impassable.

At the side of the runs for the grown fowls is seen another large grass field, reserved for the young chickens. Against the back wall of this run a number of rude sheds are erected, each covering a coop, as represented in Fig. O, page 46.

The arrangements of the ground-floor of the poultry-house are shown in plan by Fig. 20, and in perspective by Figs. 21, 24 and 25.

Fig. 21. The Hatching-room.

No. 1 on the plan, and Fig. 21 in perspective, represent the hatching-room, which is at one end of the building, and is very ingeniously contrived. The nests P are arranged on a double dresser, running round three sides of the room, and consist of wicker baskets of an oblong square form, made larger at the top than the bottom, in order better to accommodate the hen's head and tail. Each basket has a cover, and a small ring for affixing a label, to denote the date of hatching. A table with drawer, a thermometer, registry-book, with writing materials, and a small cupboard, complete the interior furniture of this room.

On the opposite side of the room to the nests, and *outside* the wall, are two tiers of coops for feeding the hens. The construction of these feeding coops, which measure sixteen inches wide by eighteen long, is more clearly shown by Fig. 22, D being a side section, and E a front view. They open at the rear into the hatching-room by trap-doors, built in the wall, and in front, on ledges. The food and water are supplied in two earthen pans, to which the hens get access by thrusting their heads through

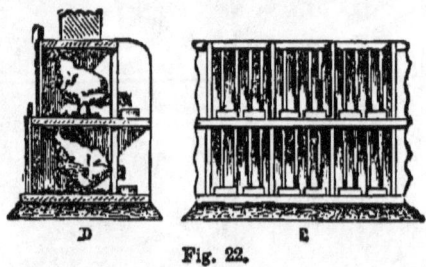

Fig. 22.

the bars. The partitions between the coops project beyond the bars, so that the hens cannot see each other whilst feeding.

The management of the hatching-room is easily understood. Each hen is taken in turn from her basket, and put through the trap-door into a coop until all are occupied, the pans having been replenished previously with food and water. They are put back in the same order as they were taken out, the attendant never leaving the room, except to clean out the coops and replenish the feeding vessels, should there be more hens than the number of coops will contain at one time. Thus all is conducted without noise or disturbance.

When hatched and strong, the hen and her brood are conveyed to the rearing-field in the quietest manner, without even taking them out of the nest, by slipping the hatching-basket (as already noticed, of a taper shape), into an iron ring furnished with handles as shown in Fig. 23.

THE KITCHEN.

Fig. 23.

No. 2 on the plan is the kitchen, shown in perspective by Fig. 24. This contains a furnace, F, with two copper boilers for cooking the roots and vegetables, a dresser, T, and the necessary

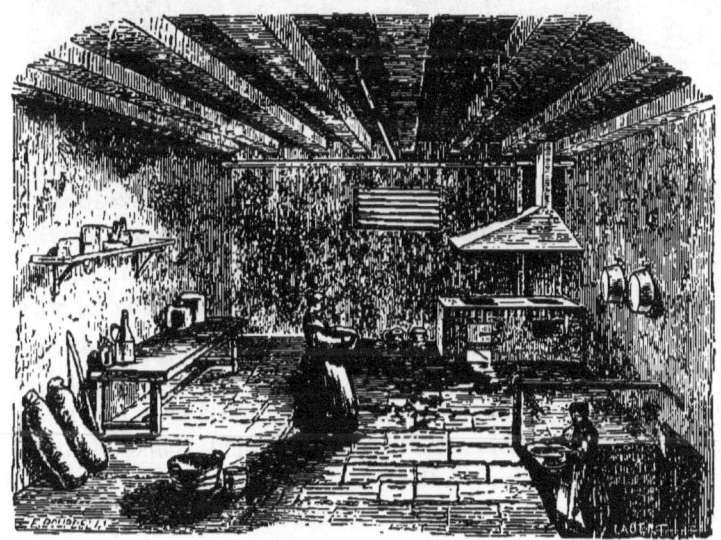

Fig. 24. The Kitchen.

shelves and utensils. In one corner is a staircase, E, leading to the cellar below, in which the potatoes and vegetables are stored. Pipes from over the furnace convey warm air to the hen-houses above, and to the hatching-room, when required.

Fig. 25 and No. 3 on the plan show the arrangements of the store-room, which contains the stock of meal and grain, in bins carefully designed for its good preservation. The corn-bins are

Fig. 25. The Store-room.

shown on a larger scale by Fig. 26, and are the invention of M. Audeod. The framework, F, is of wood, the sides of wire gauze, properly supported by additional wooden stays, T. Inside these is also a ventilating chimney, similarly constructed of wire gauze, on a wooden frame, which passes through both the lid and bottom of the bin, and maintains a draught of air through the centre of the mass, whose exterior is also ventilated through the gauze sides. The bottom is formed of a double slope, slanting like a shallow trough from the sides, A B, to the middle line, C D, and the trough also sloping lengthways from back to front. At the lowest point is a shallow spout, D, to which access is afforded by a shutter; and it will be readily seen that the bin will empty itself to the very last grain.

The Audeod corn-bin deserves to be adopted in all large poultry establishments. From the free ventilation provided, the

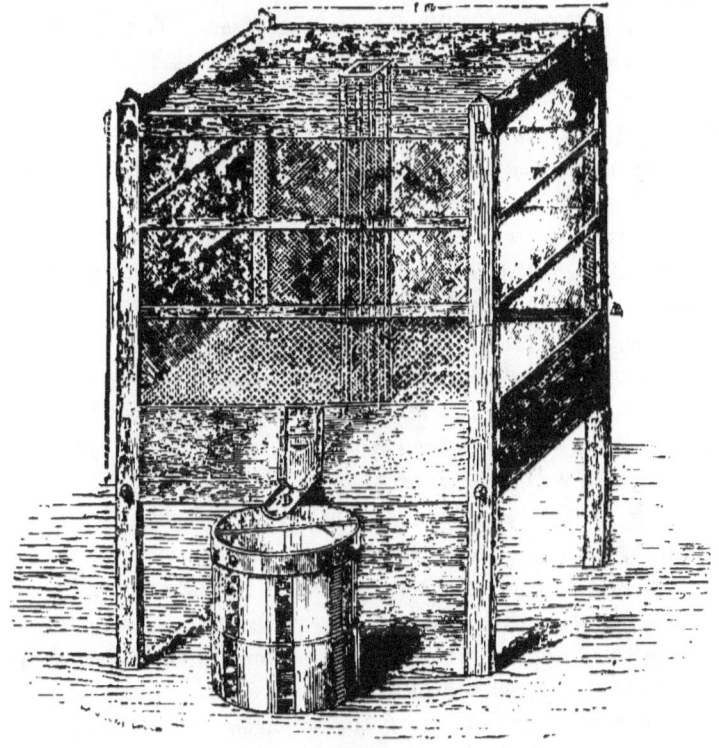

Fig. 26. The Audeod Corn Bin.

grain—however long kept—never becomes musty, but is preserved in a sound state, and the bottom, or stalest portion, is always used first. The elevation on legs is also not only convenient for delivering the corn into the receiving vessels, but secures it from the attacks of vermin.

The bins at Belair contain nearly thirty bushels each, and

five of them, c c (Fig. 20), are ranged in a line at one end of the room, lozenge fashion, in order that air may have free access to them. At the other end, one each side of the door, are two large chests, B B, for meal and bran. These cannot of course be made of gauze, and are best of sheet-iron. They have, however, gauze covers, and are inclined at the bottom like the corn-bins, so that the least aërated portion is first used. The necessary measures and vessels complete the furniture of this room.

No. 4 on the plan (Fig. 20) represents the egg-room, of which a view is unnecessary, as it is simply furnished round the walls with shelves, H, on which are placed the oblong square boxes in which the eggs are packed. Each box has marked upon it its date, which, with the date on which it leaves the establishment, is entered in a registry book. A separate corner is appropriated to eggs for hatching.

A separate building is devoted to fattening purposes, fitted up all round the interior with tiers of cages, each large enough to contain one bird. The fowls are either crammed by hand, or by a machine which has been recently invented for that purpose, but of which we cannot approve.

Such is the establishment at Belair, described by the authority already mentioned as one of the most complete and perfect of its kind in France, and to have been conducted for several years with great success. This being so, any critical remarks may appear invidious, but we must make a few observations respecting modifications which we think desirable.

We confess to not liking the arrangements for hatching. To set the hens in baskets on shelves may perhaps be inevitable in large establishments, but the birds should certainly have more room to stretch their legs when off the nest than is afforded by a pen a few inches square, and it is also needful they should have access to a dust-bath, or they are tormented by vermin to an intolerable degree. It would be better to give up more space to the hatching department, so as to give each pair of hens a *small* yard, and set them on the ground. For instance, twenty pens, 10 feet by 3 feet, would each contain two nests at one end, would give room for exercise and ablution, and would accommodate forty hens in a space of only 30 feet by 20. This would be amply sufficient to hatch 3,000 chickens per annum, and they will be of much stronger constitution than on the plan, ingenious as it is, which we have described.

Neither do we consider an acre and a half of run the most *really* economical allowance for 330 fowls, whilst we should also recommend the keeping of them in flocks of lesser number. It is true that by the quarterly digging of the runs much evil is prevented; but by keeping say 120 fowls on an acre this would be dispensed with, and the additional rent would be more than compensated by economy of food and saving of labour.

Lastly, we consider it a very unadvisable plan to select the eggs for hatching from even the finest of those laid by the general stock. It is far better, from amongst the large mass of chickens reared annually, to select the very finest specimens, and reserve them in pens of, say, one cock to from four to six hens, for breeding alone. Many advantages will be thus secured. In the first place, *all* the eggs will for certain be of first class quality, and well fecundated. The cocks in the general runs may also be reduced to about one in twenty, or even dispensed with; thus sending more to market, and saving their food. And lastly, a share of the honours of exhibition may be secured, and sums not to be despised realised by selling at high prices to amateurs. At the same time, these select yards must not be allowed to degenerate into mere pens for breeding "fancy fowls," as will be the case if not watched. They must be *mainly* regarded as the sources of supply to the general yards, and will then be found a valuable addition to the arrangements at Belair.

For the scale of that establishment, say 1,200 laying hens, we do not think its arrangements can be further improved, save that for the small detached shelters over the coops in the chicken nursery, it would be far better to substitute one long and spacious shed. We should ourselves also prefer the hen-houses on the ground-floor, in which case the kitchen, store-room, &c., might be placed behind; but these are merely matters of opinion and detail.

But on a larger scale some further modifications will be desirable, if only for the simple reason that the triangular shape of the runs will be very inconvenient if multiplied; whilst, if rectangular, as they must be the width of the houses, they would be nearly as awkward from their length and narrowness. It is needful to mention this, as we have a strong conviction that with less than 10,000 fowls there is not sufficient return to be worth the attention of the English capitalist. And whether there be an adequate market for the produce of such a number must in all cases be carefully ascertained before such an undertaking be engaged in. This much being taken for granted, we would make the following suggestions respecting the formation of a chicken farm.

Of the 10,000 fowls we would reckon 400 as the breeding stock, and 9,600 as laying, or ordinary stock, divided into 80 flocks of 120 each. These flocks should each have an acre of run; 15 acres more would be required for the chicken run or nursery; 3 acres for the breeding yards, and the remainder for hatching runs and buildings, pigs, &c., &c. In all, 100 acres.

The simplest and best arrangement would be to have the grass runs, say 80 feet by 550 feet, with a house or close shed 80 feet by 4 feet at one end, provided along its whole length with traps, for the fowls to enter. One long perch will then roost all the birds, and the nests will also be contained in a single row. In front of the house should be a shed extending about 20 feet, and floored with hard gravel or asphalt, under which the food will be thrown, and to which the birds can retreat. The runs should be side by side, and two rows of houses arranged back to back, with a passage between, into which their doors open. This passage should have a sky-light roof, and the houses be only fronted into it with netting; this part of the arrangement being like that of Mr. Lane's establishment, figured at page 65, only that each house is much longer. The nests should similarly be reached by trap-doors from the passage, which should be traversed by a railway-truck to collect the eggs and manure.

By such an arrangement, all the needful operations will be conducted with the least possible labour.

The conditions of health, fecundity, and profit will not differ from those enunciated in the first section of this work. But in a large concern all operations will range themselves into five great divisions : the breeding-yards, the hatching-pens or rooms, the chicken-nursery or rearing runs, the ordinary stock-yard, and the fattening pens. These must be arranged in any way that will best secure economy of labour and *effective* supervision.

As much machinery as possible should be employed in preparing the food, and to work these a small steam-engine will be found very economical, whilst it may be made auxiliary to cooking purposes.

Great care must be taken that the land is well drained, and, if possible, slightly sloping to the south. A light, dry soil is also very desirable, but good drainage will overcome great difficulties in this respect.

The selection of breeds is of the very utmost importance. With a good market for both eggs and fowls, we would recommend one-fourth Dark Brahmas, one-fourth Dorkings, one-fourth Houdans, and one-fourth a cross between the three, obtained by first mating the largest Brahma hens with a Dorking cock, and then breeding from the progeny with the largest Houdan cocks that can be procured. Of this cross we cannot speak too highly, as admirable chickens, thus bred, may be sent to market at ten weeks old—an earlier period than is possible with any other fowls we know. All the runs, except the Dorkings, will yield an abundance of eggs, and that breed will be most valuable for table fowls, and also as mothers. If another breed be desired, La Flèche should be selected, for the sake of their fine large eggs, combined with good and heavy table qualities. Except in very favourable situations, Crèves are too delicate to be remunerative.

The profitable disposal of the manure should be especially studied, and for this reason we should strongly recommend some measure of farming operations to be carried on in combination. A number of pigs should likewise be kept, as they may be fattened on what the fowls refuse. Or ducks will also make capital "save-alls."

We are reluctant to enter into figures, we have seen so many visionary and delusive statements; but we know that *some* data, however rough, will be expected. It is only as such that we offer the following; and if our figures do not show three hundred per cent. as the probable profit, it is because they are based upon some attempt, however rough, at calculation from actual *facts*, not upon the sanguine theories of persons totally ignorant of fowls. We shall still suppose an establishment of 10,000 birds.

For *capital* we would estimate—

10,000 fowls, at 2s. each	£1,000
Buildings, Fittings, Engine, Plant and Utensils, including 2 horses and carts	1,350
One month's food	150
Spare cash working capital	500
Say total capital	£3,000

Our plan of commencing would be to purchase first simply 400 first-class breeding birds at an *average* of about 20s. each (some would be much more than this). The stock for the second year would then cost less than £1,000, but there would be little to spare for sales.

Our estimate for the *working* of such an establishment is based upon the fact, that of all the breeds mentioned above except Dorkings, 150 eggs per annum may be obtained from each hen. Including them the average will be 140 all round, or with the cocks say 130. On the large scale, we are also satisfied that the keep of a fowl will not exceed 3s. per annum, and from these facts we are justified in reckoning every fowl in the yard as representing a gross profit (including the manure) over and above her food, of 4s. per annum, leaving all other expenses to be deducted. A rough estimate may then stand thus—

RECEIPTS.		EXPENSES.	
Gross profit over food from 9,600 stock fowls, at 4s. per annum each	£1,920	Rent—100 acres, at 40s.	£200
		Taxes	40
		Interest on capital of £3,000 at 5 per cent.	150
		Wages—2 men with their families	200
		Horse keep	60
		Fuel and attendance for engine	100
		Gross balance of profit	1,170
	£1,920		£1,920

In this estimate nothing is allowed for *renewing* the stock, because all the fowls, which should never be allowed to become *old*, can be sold when fatted for more than they actually cost as delivered from the breeding-yards.

There will be other items of expense which cannot be set down. Railway carriage is difficult to estimate, and will affect profit; there is also wear and tear to allow for. But on the other hand, the above balance-sheet represents the profit of the laying stock alone, and a gross profit of at least equal amount will be derived from the dead stock sent to market from the breeding-yards. Of this we give no details, as the returns from chickens sold at ten to eighteen weeks old—and they should not be older—may be easily estimated. In the main, therefore, the above figures will be found sufficient; and if they show a somewhat more moderate return than preceding writers have promised, they are at least likely to be realised, and certainly—making the sole and all important stipulation of *a market*—offer sound inducements to the enterprising capitalist.

It is, however, to the *farmer* that poultry-breeding on a large scale more especially commends itself; and it may be pursued most successfully on either of two quite distinct systems. A large number may be kept all through the year, and a portion of the farm—say one-fourth—permanently appropriated in regular rotation to their use, the fowls being removed to fresh ground every year. Or, on the other hand, a moderate breeding-stock only may be permanently retained, but a large number of chickens reared from them every season, which should be sent to the fields as soon as cropped, in travelling houses mounted on wheels. There they will speedily get fat at very little expense, and may be killed off for the market. The first plan is most suitable for large farmers with good business and administrative capacity; the last will be best adapted for smaller holdings. But either system will not only yield a handsome profit in itself, but greatly benefit the other produce; both by manuring the ground, and by removing myriads of worms and insects very injurious to the growing crops. Indeed, considering the ravages yearly committed on every farm by these tiny pests, it is to us most astonishing that, instead of the bungling methods of extermination at present employed, the farmers of England do not have recourse to the philosophical and lucrative remedy which nature has provided.

The choice of breeds will be generally as already mentioned, but will vary with circumstances. Dorkings should not be kept when eggs are the principal object; nor Brahmas when dead poultry is the end in view. If only one breed is desired, Houdans will be best, with a few Brahma hens for hatching and crossing.

It is on the farm poultry ought to be *most* profitable; and, in such circumstances, we consider every well-chosen stock-fowl should represent a clear profit of five shillings per annum; whilst we are quite sure chickens will yield a much heavier weight of meat for the same outlay than any other stock whatever. The time is fast approaching when this will be generally recognised; and then, and not before, will poultry-breeding occupy its legitimate position in the general economy of agriculture.

To contribute in some slight degree towards this result, has been one object of the preceding pages.

INDEX.

Accident to Eggs, How to act in case of, 40
Age to Breed from, The Best, 81, 82
Anconas, 136
Andalusian, The, 136
Animal Food necessary for Fowls, 27
——, Caution against over-use of, 28
——, How to supply, 28
Apoplexy, Treatment for, 57, 58
Artificial Hatching, 203, 204
——, Danger of Concussion in, 209
——, The Eggs used in, 209
——, Hints about, 209
——, The Sprinkling of the Eggs in, 209, 210
——, The Success of, 208, 209
——, The Supply of Heat for, 211, 212
——, The Temperature required for, 210, 211
"Artificial Mother," Description of The, 214—216
——, Schröder's, 216
Artificially-reared Chickens, Food for, 217, 218
Artificial Rearing, 213
——, Points to be attended to in, 218
——, Importance of Cleanliness in, 217
——, Value of, 218
——, Directions as to the Weather in, 215—217
Artificial Selection, Evils of too great, 75
Audeod Corn-bin, The, 234, 235

Aylesbury Ducks, The Bills of, 192, 193
——, Description of, 191, 192
——, Fattening of, 192
—— as Layers, 192

Balance-sheet for the Poultry-Farmer, A, 240—242
Bantams, 162, 163
——, Black, 165
Bantam Chickens, Treatment of, 167
Bantams, Cochin, 166
Bantam Eggs, 167
Bantams, Game, 165
——, Japanese, 166, 167
——, Laced, 77, 78
——, Nankin, 166
——, Pekin, 166
——, Sebright, Description of, 163, 164
———, The Two Varieties of, 164
——, Usefulness of, 167
——, White, 165, 166
Barley as Food, The Use of, 27
Barn-door Fowls, 17
Belair Establishment, Remarks on the, 236—238
—— Poultry Farm, Description of the, 224—236
Black-breasted Reds, The, 121
Black-crested White Polands, 147
Black Game Fowls, The, 122
Black Hamburgs, 78, 144
Black-winged Pea-fowl, 183
Box for Carriage of Hen and Chickens, 47

R

Boyle's, Description of Mr., Establishment, 67—69
Brahmas, The, 15—17
———, The Breed of, 105, 106
———, The Breeding of, 110, 111
———, The Dark, as Breeding-fowls, 19
———, The Colour of, 111—113
———, The Comb of, 106
———, The Cross of, 107
———, The Dark, Description of, 108—110
———, The Form of, 111
———, The Light, Description of, 107
———, The Merits of, 115
——— as Mothers, 37
———, The Dark, Mr. Boyle's Opinion of, 108—110
——— as Table-fowls, 115
Breda, The, 160, 161
———, Merits of, 161
Breed from, The Best Age to, 81, 82
Breed, Test for the Accidental and the True, 76
Breeder, Errors on the part of the, 78
———, The Power of the, 78
Breeding, What Attentive, might do, 79, 80
Breeding-season, Shelter during the, 45—47
Breeds, Selection of, for Chicken-yard, 239, 240
Breeds for the Poultry-yard, Choice of, 20
Bresse, La, 161
Brindley's Incubator, 206, 207
Brown-Reds, The, 120, 121
Buckwheat as Food, 27
——— as Food prior to Shows, 95
Buenos Ayres Black Ducks, 195, 196
Buff Polands, 149

Call Ducks, 195
Canada Geese, 198
Carbonnier's Incubator, 204—206
Cats, How to prevent the Inroads of, 48, 49
Cayuga Ducks, 196
Chamois Polands, 149
Charny, Description of the Establishment at, 224—226

Charny, Remarks on the Establishment at, 236—238
Chick, Separation of the, from the Shell, 43
———, How to act when the, sticks to Shell, 43
Chicken-farm, Formation of, 238, 239
———, Selection of Breeds for, 239, 240
Chickens, Importance of Cleanliness to, 48
———, Benefit to, of Cooping near Grass, 47, 48
———, Dark, 123
———, Evil of making, Drink when Hatched, 45
———, Exhibition, Treatment of, when Hatched, 86—89
———, How often, should be Fed, 50
———, Food for, as they grow older, 50
———, Food for, in Cold Weather, 51
———, How the Hen can escape the Worry of, 85
———, The Best Condition of, for Home Use, 54
———, Light Game, 123
———, First Meal of, 45
———, When, are ready for Exhibition, 90
———, Prize, The Best Food for, 86—89
———, When, should Roost, 89
———, Striped, 123
———, Treatment of, in Cold Weather, 51
Chicks, The Best Food for, 49, 50
Chinese Geese, 197, 198
Cleanliness in Artificial Rearing, 217
——— in the Roosting-house, 7
Cochin Bantams, 166
Cochins, About, 101, 102
———, General Appearance of, 103, 104
———, Build of, 102, 103
———, Characteristics of, 102
———, Colours of, 103, 104
———, Effect of Crossing on, 77
———, Defects of, 105
———, Diseases to which, are liable, 105
———, Merits of, 104
——— as Mothers, 37
———, Relative Value of, 105

Cochins, Silky, 170
——, Weight of, 102
Coloured Dorking, Effect of Crossing on, 77
Colonel Stuart Wortley's Incubator, 212, 213
Columbian, The, 136
Common Pheasant, The, 188
Concussion, Danger of, in Artificial Hatching, 209
Condition, to preserve Fowls in good, 84
——, What is really good, 91
Construction of the Fowl-house, 3, 4
Corn-bin, The Audeod, 234, 235
Cramp, Treatment for, 56
Creepers, 167, 168
Crèvecœurs, 152, 153
——, Faults of, 154
——, Merits of, 153, 154
Crossing and Selection Combined, Results of, 77, 78
Crossing, Effect of, on Cochins, 77
—————, on Coloured Dorking, 77
—————, on Game Fowls, 77
—————, on Surrey Fowl, 77
——, Examples of, 77, 78
——, The Principle and Advantage of, 76
Cygnets, Treatment of, 200

Dark Brahmas as Breeding-fowls, 19
——, Description of, 108—110
——, Mr. Boyle's Opinion on, 108—110
Dark Game Chickens, 123
Dark Greys, The, 122
Deodorisers, The Best, 9, 10
Diarrhœa, Treatment for, 59
Disease, The Best Cure for, 55
——, How to Prevent, 55
——, General Symptoms of, 59
——, Treatment upon Appearance of, 59, 60
Domestic Poultry-keeping, Profits from, 32, 33
Dorkings as Breeding-fowls, 19
Dorking, the Coloured, Effect of Crossing on, 77

Dorkings, Diseases to which, are liable, 130
——, General Description of, 126, 127
——, Effects of Inter-breeding on, 129
—— as Layers, 130
—— as Mothers, 37
——, Hints about the Rearing of, 130, 131
——, The Silver-grey, 128
—— as Table-fowls, 130
——, The White, 128, 129
Douglas Mixture, Use of the, 30
Draughts in the Fowl-house, 4
Ducks, Call, 195
——, A Few Hints respecting, 196
——, Houses for, 196, 197
——, Large Black, 196
—— as Layers, 196
——, Muscovy, 195
——, The Rearing of, 190, 191
——, Use of, 191
Dumpies, 167, 168

East Indian Black Duck, 195, 196
Eggs, How to act in case of Accidents to, 40
——, The Sprinkling of the, in Artificial Hatching, 209, 210
——, The, used in Artificial Hatching, 209
——, Advantage of Distinguishing the, 32
——, Testing the Fertility of, 41, 42
—— for Setting, Importance of using Fresh, 34, 35
——, How often, should be Gathered, 32
——, The Proper Number of, for Hatching, 42
Egg-laying Stocks, The, 16
Eggs for Setting, How to Keep, 35
——, How to Pack, 86
—— for Setting, The Selection of, 34
Embden Geese, 197
Emu, The, 170
English White Swans, 199
Exhibition, When Chickens are ready for, 90
——, Proper Food during, 96, 97

248 INDEX.

Exhibition, Immediate Preparations for 93, 94
———, Treatment after, 97
Exhibition Chickens, Treatment of, when Hatched, 86—89
Expense of keeping a Poultry-farm, Estimate of, 240—242

Fancy Poultry, 17
Farm, Advantage of Rearing on a, 70
Farmer, Usefulness of Poultry to the, 242, 243
Fat, Hints as to Extra Weight and, 53, 54
Fattening, The Best Food for, 53
Fattening Hens, 52, 53
Fattening, Duration of the, Process, 53
———, Objections to, 91
———, The Secret of, Profitably, 53
Feathers, How to Dress, 32
———, Loss of, Treatment for, 58
———, Value of, 32
Feeding, Errors to be guarded against in, 23, 24
———, Importance of Judicious, 20, 21
———, Evils of Over-, 21, 22
—— of Sitting Birds, The, 37
Flèche, La, 154
———, Merits of, 155, 156
Fledging, Treatment for Bad, 56
Food, Animal, necessary for Fowls, 27
—— for Artificially-reared Chickens, 217, 218
———, Benefit of Change of, 25
———, Best, for Chicks, 49, 50
—— for Chickens in Cold Weather, 51
—— for Growing Chickens, 50
—— for Prize Chickens, 86—89
———, How to give, to the Fowl, 26, 27
—— for large number of Laying-fowls, 24, 25
—— for small number of Laying-fowls, 24
———, Best, for Morning Meal, 23
———, Best, for Evening Meal, 23
———, Rule for Regulating the Supply of, 22

Food, To Mix Soft, 26
———, Vegetable, Importance of, 28
Foods, Analysis of various, 25
Fowl-house, Construction of the, 3, 4
———, Draughts in the, 4
———, The Flooring of the, 7
———, Size of the, 5
———, Ventilation of the, 5
Fowls, Attention to be paid to, 3
———, What makes the Best, 91
———, How to keep them in Good Condition, 84
———, How Old, should be Cooked, 55
———, What to do with Old, 15, 16
———, Various, 172
———, How to tell Young, 15
Fowl-sheds, The Advantage of, 6, 7
French Breeds, The, 151
———, Features of the, 162
Frizzled Fowls, 170

Game Bantams, 165
—— Chickens, Light, 123
———————, Striped, 123
Game Fowls, The Black, 122
———, The Best Criterion for Blood in, 123
———, The Breeding of, 124, 125
———, Effect of Crossing on, 77
———, Demerits of, 125, 126
———, Description of, 118—120
———, Best Fighters among, 123, 124
———, Time for Hatching Eggs of, 125
———, Best Layers among, 124
———, Merits of, 125
———, Original Wild Varieties of, 123
Game Hens, 17
—— as Mothers, 37, 38'
Gapes, Treatment for, 57
Gardener's Friend, The, 191
Geese for Exhibition, Treatment of, 199
———, General Management of, 198
—— as Sitters, 198
Golden "Mooney" Hamburgs, 140, 141
Golden-pencilled Hamburgs, 140
Golden-spangled Hamburgs, 140, 141
Golden Pheasant, The, 188, 189
Golden-spangled Polands, 148, 149

Golden Yorkshire Pheasant-fowl, 141
Goslings, Treatment of, 198, 199
Grass, Benefit of Cooping near, 47, 48
Green Food, Great Value of, for Prize Fowls, 89
Gueldres, The, 161
Guinea-chicks, Treatment of, after Hatching, 182
Guinea-fowls as Layers, 182
———, Remarks on, 181, 182

Hamburg Cocks, Silver-pencilled, Chief Faults in, 139
——— Hens, Silver-pencilled, Chief Faults in, 140
Hamburgs, The, 16, 138, 139
———, The Black, 78, 144
———, Golden "Mooney," 140, 141
———, Golden-pencilled, 140
———, Golden-spangled, 140, 141
———, Lancashire Silver "Mooney," 142
——— as Layers, 144, 145
———, Silver-pencilled, 139
———, Silver-spangled, Breeding, 142, 143
———, Silver-spangled, Breeding, Proper Mode of, 143
Hamper, Best Form of, 95
Hatching, Hints to Buyers of Eggs for, 85, 86
———, Proper Number of Eggs for, 42
———, Preparations for, 44, 45
Hatching, Time of, 43
———, Time of, for Shows, 85
———, Treatment immediately after, 44
Hatching-run, Proper Arrangement of the, 36, 37
Heat, The Supply of, for Artificial Hatching, 211, 212
Hempseed as Food Prior to Shows, 95
Hen, Effect of, on Progeny, 83
Hens, Number of, to the Pen, 84
Houdans, 157—160
——— as Breeding-fowls, 19
———, Merits of, 160
Hybrid Pheasants, 189, 190

Incubator, Brindley's, 206, 207
———, Carbonnier's, 204—206
———, Schröder's, 207, 208
———, Jacob Graves and Sons, 212, 213
Indian Meal as Food, 26
Insect Vermin, Treatment for, 59

Japanese Bantams, 166, 167
Javan Pea-fowl, 183
Judges, their Merits and Demerits, 96
Judging, Objections to the Present System of, 78, 79

Killing, The Best Mode of, 54
———, The Various Modes of, 54

La Bresse, 161
Laced Bantams, 77, 78
La Flèche, 154
———, Merits of, 155, 156
Lancashire Silver "Mooney" Hamburgs, 142
Lane's, Description of Mr., Establishment, 64—67
Large Black Duck, 196
Laying Hens, Food for a Small Number of, 24
———, Food for a Large Number of, 24, 25
———, How to Keep, 14, 15
Leg Weakness, Treatment for, 56
Light Brahmas, Description of, 107
Light Game Chickens, 123
Lime, Value of, 30, 31
Linseed as Food prior to Shows, 94
Linton Poultry-yard, Description of the, 69

Maize as Food, 26
Malays, Description of the, 116, 117
———, Faults of the, 117
——— as Table-fowls, 117
Male, How the, affects the Progeny, 83
Manure, Use of Poultry-, 31, 32
Matching, Points to be Considered in, 92
Meal, The Best Food for Mid-day, 27
———, The Best Food for Morning, 23
———, The Best Food for Evening, 23

Meals, Number of, a Day, 23
Minorca Spanish, The, 135
Moisture, Importance of, to Setting Hens, 39
"Mooney" Golden Hamburgs, 140, 141
———, Lancashire Silver, 142
——— and Golden Yorkshire Pheasants, Show-breed from, 141, 142
Moulting, Bad, Treatment for, 56, 57
———, Treatment during, 30, 31
Muscovy or Musk Ducks, 195

Nankin Bantams, 166
Negro Fowls, 168—170
Nest, The Construction of the, 31
———, The Formation of the, 39, 40
———, Absence of Hen from the, 42, 43
———, Arrangement of, for Sitting Birds, 38

Old Fowls, How, should be Cooked, 55
———, What to do with, 15, 16
Over-feeding, Evils of, 21, 22

Park, Advantages of Rearing in a, 69
Pea-chicks, Rearing of, 184
Pea-fowl, Black-winged, 183
———, Disposition of, 183, 184
———, Javan, 183
Pea-fowls, Where, should be kept, 184
———, Remarks on, 182, 183
Pekin Bantams, 166
Pencilling, 114
Perch, Size and Position of the, 5, 6
Pheasant Chicks, Early Treatment of, 187, 188
———, Treatment of, 185, 186
Pheasants, Best Food for Adult, 186
———, Treatment of, after Breeding-season, 188
———, The Common, 188
———, Best Diet for, 186
———, Collecting the Eggs of, 187
Pheasants' Eggs, Hatching of, 187
Pheasants, To avoid the "Gapes" in, 186
———, The Golden, 188, 189

Pheasants, The Handling of, 190
———, Hybrid, 189, 190
———, Pens for Rearing, on a Large Scale, 186, 187
———, The Silver, 189
———, How to make, Sit, 185
———, Treatment of, 185
Piles, The, 122
———, The Prize, 122
———, The Worcestershire, 122
Pip, Treatment for, 59
Plucking, 54
———, Hints as to Appearance after, 55
Plumage, How to Modify the, 83
Polands, Black-Crested White, 147
———, Buff, 149
———, Chamois, 149
———, The Comb of, 145, 146
———, Great Faults in, 150, 151
———, Distinguishing Features of, 145
———, Golden-spangled, 148, 149
———, Merits of, 150
———, Precautions Necessary in Rearing, 150
———, Silver-spangled, 147, 148
———, Tenderness of, 150, 151
Poland, White-crested Black, Cocks, 146
———, White-crested Black, Hens, 146, 147
Polands, White-crested White, 147
Potatoes as Food, 26
Poultry-keeping a Matter of Business, 33
———, The Great Essential in, 9
Poultry-manure, Value of, 31, 32
Poultry, The Descent of, 71—74
———, Usefulness of, to the Farmer, 242, 243
Poultry-rearing as a Business, 221—224
Poultry-yard, Choice of Breeds for, 20
———, Plans of a, 10—13
Prize Chickens, the Food for, 86—89
———, When, should Roost, 89

Rearing in a Park, Advantages of, 69
——— Prize-fowls, Space required for, 64

INDEX. 251

Rearing on a Large Scale, Space required for, 64
Redcaps, The, 168
Reds, The Black-breasted, 121
——, The Brown, 120, 121
Relationship, To provide against, in Breeding, 82
——, Attention to be paid to Degrees of, 82
Roosting-house, Cleanliness in the, 7
Rouen Ducks, Description of, 193, 194
—— and Drakes, Difference between, 194
——, Fattening of, 194
Rouen Ducklings, Treatment of, 195
Roup, A Fact worth Knowing about, 57
——, Treatment for, 58, 59
Rumpless Fowls, 170, 172
Run, How the, should be kept, 8
——, Space required for the, 9
Russians, 172

Schröder's "Artificial Mother," 216
—— Incubator, 207, 208
Scientific Theory of Breeding, Importance of, 70, 71
Sebright Bantams, Description of, 163, 164
——, The Two Varieties of, 164
Selection, Evils of Over-great Artificial, 75
—— and Crossing Combined, Results of, 77, 78
——, Effects of, 74, 75
——, Examples of, 76, 77
Setting, Importance of using Fresh Eggs for, 34, 35
——, How to keep Eggs for, 35
——, The Selection of Eggs for, 34
—— in Winter, 40, 41
Sexes, Separation of the, 89, 90
Shell, How to act when the Chick adheres to the, 43
——, How to assist Chicks from the, 43
Silky Cochins, 170
—— Fowls, 168—170
Silver Duck-wing Greys, The, 121, 122

Silver-grey Dorkings, The, 128
Silver-pencilled Hamburgs, The, 139
Silver Pheasant, The, 189
Silver-spangled Hamburgs, Breeding, for Exhibition, 142, 143
Silver-spangled Polands, 147, 148
Silver Yorkshire Pheasants, 142
Sitting, How to Prevent, 15
Sitting Hens, Benefit of Moisture to, 39
——, Arrangement of Nests for, 38
——, Management of, in the Ordinary Nest, 36
——, Separate Provision for, 35
——, Qualifications necessary in, 37, 38
——, Protection of, against the Weather, 38, 39
Sitting-shed, Best Situation for, 39
Size, How to Increase the, 83
Snow-water, Evil Results from use of, 29, 30
Soft Egg, Treatment for, 59
Spanish, The, 16
——, The Comb of the, 132—134
——, Diseases to which the, are liable, 136, 137
——, Treatment of, before Exhibition, 138
——, The Face of the, 132, 133
——, The, as "Fancy" Fowls, 137
——, The, as Layers, 137
——, Merits of the, 137
——, The Minorca, 135
——, Importance of Purity of Race in the, 134
—— and other Delicate Breeds, Establishment for Rearing, 64—68
——, The, as Table-fowls, 137
——, The White, 135, 136
——, Description of the White-faced Black, 132, 133
Striped Game Chickens, 123
Sultans, 149, 150
Surrey Fowl, Effect of Crossing on the, 77
Swans, 199
Swan, The Black, 199
——, The English White, 199

Swans as Layers, 200
"Sweepings," Objections to use of, as Food, 27

Temperature necessary in Artificial Hatching, 210, 211
Toulouse Geese, 197
Turkey-chicks, Diseases to which, are liable, 179, 180
——, Proper Food, for, 178, 179
——, Danger of Moisture to, 176
——, Treatment of, 176
——, Treatment of, immediately after Hatching, 178
——, Danger of Bad Weather to, 180
Turkey-cock, Disposition of, 177
Turkey-hen, Disposition of, 177, 178
Turkey-hens as Layers, 177
Turkeys, The Breeding of, 176, 177
——, The Foreign Breeds of, 180, 181
Turkey's Hatching, Duration of, 178
Turkeys, Kinds of, 180
——, Remarks on the Rearing of, 175
——, Size of, 180
——, Treatment of, when Hardy, 180
——, Treatment of, immediately before Hatching, 178

Utility, The Great Advantage of, 79

Various Fowls, 172

Vegetable Food, Importance of, 28
Vegetables, Benefit of, to Chickens, 51
——, What, may be used, 28, 29
Ventilation of the Fowl-house, 5
Vermin, Precautions against, 217
——, Insect, Treatment for, 59
Vulture Hocks, 114

Water, Fresh, Importance of, 29
Water-vessel, Best Kind of, 21
Weather, Hints as to, in Artificial Rearing, 215—217
Weight. Hints as to Extra Fat and, 53, 54
Wheat as Food, 27
White Bantams, 165, 166
White-crested Black Poland Cocks, 146
—— Black Poland Hens, 146, 147
—— White Polands, 147
White Dorkings, The, 128, 129
Whites, The, 122
White Spanish, The, 135, 136
Worcestershire Piles, The, 122
Wortley's (Colonel Stuart) Incubator, 212, 213

Yellow Duck-wings, The, 122
Yorkshire Golden Pheasants and "Mooneys," Exhibition Fowls from, 141, 142
Yorkshire Silver Pheasant Fowls, 142
Young Fowls, How to tell, 15

www.ingramcontent.com/pod-product-compliance
Lightning Source LLC
Chambersburg PA
CBHW032105220426
43664CB00008B/1142